THE DAWN OF
THE DEED

THE DAWN OF THE DEED

The Prehistoric Origins of Sex

JOHN A. LONG

The University of Chicago Press
Chicago and London

The University of Chicago Press, Chicago 60637
The University of Chicago Press, Ltd., London
© 2012 by The University of Chicago
All rights reserved. Published 2012.
Printed in the United States of America

21 20 19 18 17 16 15 14 13 12   1 2 3 4 5

ISBN-13: 978-0-226-49254-4 (cloth)
ISBN-13: 978-0-226-00211-8 (e-book)
ISBN-10: 0-226-49254-0 (cloth)

Library of Congress Cataloging-in-Publication Data

Long, John A., 1957–
  [Hung like an Argentine Duck]
  The dawn of the deed : the prehistoric origins of sex / John A. Long.
    pages ; cm
  Originally published under the title: Hung like an Argentine Duck: a journey
back in time to the origins of sexual intimacy
  Includes bibliographical references and index.
  ISBN 978-0-226-49254-4 (cloth : alkaline paper) —
  ISBN 0-226-49254-0 (cloth : alkaline paper) —
  ISBN 978-0-226-00211-8 (e-book) 1. Sex (Biology) 2. Evolution (Biology)
3. Sexual behavior in animals. 4. Sexual instinct. 5. Sexual selection in animals.
I. Title.
QH481.L66 2012
573.6'374—dc23
2012007439

For Heather:
my best friend, beloved wife,
and fun-loving, fossil-hunting companion

CONTENTS

FOREWORD

Born and raised in Buenos Aires, Argentina, I speak with authority when I say that Argentines are going to find this new book irresistible. Adding to their pride as the self-proclaimed country with the most beautiful women (who even charmed the young Charles Darwin during his famed journey on HMS *Beagle*) and the most delicious beef, the attributes of the Argentine duck *Oxyura vitatta* will undoubtedly augment Argentinians' *T. rex* appetite for superlatives. I say this with delight because this curious duck – whose penis is as long as its body – gives us another startling example of the wonders of nature and makes us ponder the evolutionary origin of complex biological structures.

At a recent lecture at UCLA on the early evolutionary history of birds (or we may say the late evolution of dinosaurs), I was somewhat astounded by a totally unexpected question. After my talk a young student seemed neither interested in the remarkable menagerie that flew over the heads of dinosaurs nor in how animals close to the fearsome *T. rex* became transformed into the gracious hummingbirds that visit our gardens. Instead, all she wanted to know was how *Stegosaurus* did it! Taken aback, I managed barely but rather humorously to reply, 'With difficulty, I suppose' — as if having an enormous chubby body, four rather short legs, and a back lined with sharp plates and spikes were altogether a good recipe for easy sex. Upon recovering I went on to describe how dinosaurs presumably had penises. Also, how the size of these animals, and in some instances, their elaborate body gear, was suggestive of these sex organs being rather large — although no-one has yet discovered a dinosaur with a fossilized penis. My initial reaction notwithstanding, the fossil record tells us that stegosaurs were a highly successful group of dinosaurs which diversified over more than 50 million years — a feat that undoubtedly underscores how copulation was probably not as difficult as it may appear. Such flourishing evolutionary history also stresses the fact that in nature, size does indeed matter! We evolutionary biologists are very much aware that size is an undeniable driving force in evolution, and that during the history of a lineage, changes in size are often either preceded or followed by modifications in structure and/or function. We can only speculate and wonder about the changes in the shape and size of reproductive organs, as well as sexual behaviors that may have taken place as dinosaurs such as stegosaurs evolved their gargantuan sizes and elaborate body gear from substantially smaller and less ornate forebears.

Yet, as John Long's new book so vividly illustrates, the tantalizing image of an intimate moment between two stegosaurs is merely a drop in the pond of evolutionary experimentation in sex. The careful examination of a rich diversity of sexual strategies contained in these pages makes us reflect on how nature speaks with a remarkable degree of tolerance on a subject on which we often remain intolerant. This tour de force on the plurality of sex in nature should drive home the realization that when it comes to our own sexuality, there is little we should regard as unnatural.

Luis M. Chiappe
Dinosaur Institute
Natural History Museum of Los Angeles County

PREFACE

Sex, Death and Evolution

The great English Romantic poet Percy Bysshe Shelley, perhaps the most accomplished lyricist of all time, died on 8 July 1822 in a tragic accident when an unexpected storm sunk his schooner, the *Don Juan*, also taking the lives of his lover, Edward Elliker Williams, and boat boy, Charles Vivian. Two years later his poem 'The Boat on the Serchio', believed to be Shelley's tribute to Williams, was published. His metaphorical reference to the intimate act of orgasm – 'The wave that died the death which lovers love' – captures the essential poetic beauty of the experience of orgasm. Shelley's words also bring to mind the French concept of 'La Petite Mort', which embodies

ιe orgasm. Death and

ld regularly gather to

here wasn't television

egular observers of

that the poor man's

tied by an erection.

isually immediately

...medical science has clarified

that loss of oxygen to the head stimulates erotic feelings. Positron emission tomography (PET) scans of the human brain during orgasm show a depletion of blood supply to the specific area (the orbitotemporal region) that directs behavior. Dieter Vaitl, of the University of Giessen, and his colleagues reviewed the activity of the brain during altered states of consciousness. They relate how the brain goes through a momentary loss of consciousness at the time of orgasm, in effect a miniature death at the time of ejaculation.

Add to this mix several famous cases of erotic asphyxiation where sexual partners (or one person alone) try to almost strangle one another to deplete oxygen and heighten orgasmic intensity. Although Australian rock star Michael Hutchence from the band INXS may have been a high-profile example of this, none could be more infamous than the case of a Japanese woman named Sada Abe. Abe's fame arose from an incident in which she killed her lover, Kichizo Ishida, on 18 May 1936 by wrapping her *obi* (or traditional sash) around his neck and strangling him. This came only days after heated sessions in which the pair had repeatedly used their belts to cut off each other's breathing during orgasm to increase its intensity. However, after killing her lover, Abe cut off his penis and testicles and carried them around in her handbag for the next three days, until she was arrested. Perhaps this is taking a good time just a bit too far. Nonetheless, the tale demonstrates that we humans have a complex and highly variable sexuality. Sometimes its extreme pursuits lead to our demise, much like the fate of the male praying mantis. But in the vast majority of cases it results in the increase of our species population and broadens our genetic variability, much as it is supposed to.

The ultimate link between sex and death is evolution. Although a complex process, in essence evolution is the gradual transformation from one species to another over time as creatures adapt to environmental changes, or develop internal changes that enhance their ability to reproduce. In Charles Darwin's day evolution was mainly seen as 'survival of the fittest', although he did recognize sexual selection as an important driver of evolution. Today no biologist would doubt that the complex reproductive behavior and physiology of organisms have evolved over time to elaborate levels. The great diversity of life on our planet is estimated at between 8 and 10 million species, with less than 5 per cent of these formally described (although 99 per cent of all species are insects, other invertebrates, bacteria and microorganisms). This highly successful diversification of life is indeed due mainly to efficient and sustainable methods of reproduction. Life first emerged from the primeval strands of self-replicating RNA that appeared on the Earth around 3.8 billion years ago to ultimately evolve and culminate in us humans, in all our guises and with all our fickle ways.

We humans have a long and complex evolution story, and it can start from many arbitrary beginnings. I prefer to begin it with the first backboned animals some 500 million years ago, and have argued strongly that most of the human body plan was already well established in fish of the Devonian age about 360 million years ago. At that time the most advanced vertebrates had skulls containing several of the same bones we possess (for example, frontals and parietals, dentaries, maxillae), a rigid backbone with vertebrae and ribs, fore limbs and hind limbs containing the same set of bones as exist in our arms (humerus, ulna, radius) and legs (femur, tibia, fibula). These advanced lobe-finned fishes (fishes having robust paired fins supported by a strong series of bones) had multi-chambered hearts, breathed air to some degree, and possessed lungs and all the necessary modifications to invade land – which they did soon after developing fingers and toes to become early 'tetrapods'. Tetrapods include all amphibians, reptiles, mammals and birds – creatures that originally had four limbs even though they may have evolved differently, losing 'fingers' or 'toes' or even all limbs (for example snakes). Once these early tetrapods left the water and invaded land, much of the subsequent evolutionary changes were subtle finetuning of an existing body pattern rather than quantum leaps.

Thus in the evolution of mammals from reptiles, and of primates from early mammals, there are only minor changes in the skeletal patterns and ratios. The genus *Homo*, to which we humans belong, appeared on the planet some 2.4 million years ago in Africa, with modern humans, *Homo sapiens*, possibly emerging between

300,000 and 200,000 years ago. At one stage the entire population of humanity narrowed to a bottleneck. Studies of our DNA (or deoxyribonucleic acid, the unique chemical building block of life that is characteristic for every living thing) suggest that around one million years ago there were no more than about 20,000 modern humans alive, from which all seven billion Earth dwellers today are descended. Reproduction has no doubt played a major role in our species' success in taking over the planet.

Almost all of what we know about sex in humans and animals derives from behavioral observation, anatomical dissection, genetic testing and laboratory experimentation. All animals are connected through their shared DNA, so the study of their evolution provides interesting clues as to how and when certain sexual behaviors might have developed. We can also use the degree of similarity in a species' DNA to derive estimates of the evolutionary pathways back through time and identify nodes (or times) when major lineages of animals or plants diverged from one another. For example, the 98 per cent similarity in human and chimpanzee DNA can be viewed beside molecular rates of mutation to gauge an approximate time for when we diverged from the chimps. These indicate that the two species diverged around six to seven million years ago from a common ancestor. To test such predictions we can also go out into the field, find fossil remains and date them using radiometric methods that give us fairly accurate times for when the fossils were buried. In most cases, fossils are bones or impressions of the remains of once-living creatures. In rare cases they can also have soft tissues preserved, and in extremely rare cases they can even give us a window into reconstructing ancient behavior.

The idea for this book came about because a group of us paleontologists – scientists who study fossils – discovered some of these extremely rare fossils, which have in turn opened a spectacular window into understanding the very beginnings of sexual intimacy in our distant backboned ancestors. Using this research I aim to present the story of how sexual behavior first evolved in our lineage, the vertebrates, through the tale of our remarkable discoveries, both in the field and laboratory, and how we deduced our conclusions. This occupies the first seven chapters of the book, and while this part is largely an autobiographical narrative, it also attempts to explain how we 'do' science and how new discoveries are officially published and publicized.

In the remaining chapters of the book I review what we actually know of sexual anatomy and behavior in other fossils of different ages, from 560-million-year-old-algae and 100-million-year-old 100-ton dinosaurs to mammals and humans. The information *not* told by fossils, as revealed by research on sperm competition and evolutionary developmental biology, is also touched upon in the last two chapters to complete the picture.

The word 'penis' is used by some biologists strictly to denote the mammalian male organ, but I use the word throughout this book in the broader sense to define any male reproductive structure used to transfer sperm into a female. Of course, correct technical terms for these structures are used as well when they have another name (for example, the aedeagus in insects, or claspers in sharks).

Tidbits of information about some of the most unusual reproductive behavior known in closely related living creatures show the breadth and depth of animal sexual behavior, and allow us to speculate on how some of these prehistoric creatures might have mated. And the subject of genital organ evolution is finally explored with a view to determining both the commonalities between reproductive structures through evolution and why there is such widespread variation in the mating organs of animals today.

In Jared Diamond's classic book *Why Is Sex Fun?* the comparison is made between humans and other primates (gorillas, chimps, orang-utans, monkeys etc). Diamond convinces us that much of our common human sexual behavior is largely explained in terms of evolved social behaviors observed in our closest living relatives, the great apes, coupled with physiological evolution concerning visible versus hidden ovulation signaling in females. In recent times, all kinds of unusual sexual behaviors once thought to be unique to humans (because they don't appear to enhance procreativity) have been observed in a wide variety of wild creatures. From gay penguins, lesbian ostriches, necrophiliac snakes, gang-banging echidnas to fellating fruit bats, it's all described herein, based on recently published scientific papers.

This book doesn't attempt to cover the entire spectrum of animal sexual behavior as it focuses more on what we can tell from fossils and how that translates to the relevant groups of animals alive today (for a general overview of animal sexual behavior I defer to Susan Windybank's *Wild Sex*). The broader coverage of the evolution of sexual intimacy in the animal kingdom provided herein aims to demonstrate that the gap between human and animal behavior is not as large as many people might have thought.

Ultimately, I hope that this book enables you to understand another fascinating perspective on our own evolution, particularly our sexuality. Like an eagle up high looking down, such breathtaking views of evolution can only be appreciated by taking a very big step backwards from the primates to the primeval beginnings of our vertebrate line. Without wishing to sound like a Peeping Tom, I hope you enjoy the view.

John Long
January 2011

1

The Machismo of the Argentine Duck

Time is a substance from which I am made.
Time is a river which carries me along, but I am a
river; it is a tiger that devours me, but I am a tiger;
it is a fire that consumes me, but I am the fire.

Jorge Luis Borges
(Argentine poet and essayist, 1899–1986)

The male Argentine duck (*Oxyura vittata*) sports a penis as long as its entire body. The longest one measured was 1.3 feet long (42.5 centimeters) on an average-sized duck. It was discovered by a team of zoologists led by Dr Kevin McCracken of the Institute of Arctic Biology and the Department of Biology and Wildlife in Alaska.

The corkscrew-shaped penis of this duck is the largest known, relative to an animal's body size, in the entire vertebrate world. Even the mighty blue whale, spanning 100 feet (30 meters) in length with a penis extending 8 feet (2.5 meters) outside its body, doesn't come close to this little duck in overall body size comparison. (Of course, one must bear in mind the obvious difficulties involved in attempting to take measurements of an erect blue whale penis, or 'dork'. They only come out during the intensely emotional time of mating, and there are few brave enough to saddle up next to one with tape measure in hand whilst the male is desperately trying to implant his dork into the vagina of his massive mate. Needless to say these figures are only estimations from nearby observers.)

For your general interest, and we really should get this important fact out of the way at the start of the book, the human male penis ranges between 5.1 and 5.9 inches (13 and 15 centimeters) in length (for 95 per cent of all males measured), with the largest human penis size ever officially recorded being 13.5 inches (34.3 centimeters) long by 6.2 inches (15.7 centimeters) in circumference. The honor of taking this particular measurement fell to the famous American obstetrician and gynaecological researcher Dr Robert Latau Dickinson (and yes, I do acknowledge a certain irony in his surname). Dickinson (1861–1950) was not only a prolific medical scientist and surgeon but an enthusiastic public health educator, as well as a talented author and artist. He drew, painted and made sculptures of the sometimes remarkable things he saw on his travels. Many of his works were published and some are preserved in the Library of Congress in Washington DC. His big claim to fame, though, is that he was the first surgeon to introduce many of the standard gynaecological practices which are nowadays routine, such as tying off the umbilical cord of newborn babies before severing the cord after childbirth. He was also one of the first medical researchers to obtain detailed sexual histories of his patients, often using his artistic talents to make accurate drawings of many of his patients' genitalia. In his lifetime he recorded some 5200 individual case histories, so he was clearly the top expert on genital size and shape variation for his day. His measurement of the largest known human penis was taken around 1900 (owner unknown), so we can rest assured it did not involve any surgical augmentation such as might occur today.

Humans, in fact, hold the record for penis size amongst our anthropoid cousins, the great apes. The male gorilla has an erect penis size of only 1.5 inches

(4 centimeters) and an orang-utan is only slightly larger, but the chimpanzee's tackle is about twice that size. To be 'hung like a gorilla' would be considered a grave insult indeed in some parts of Africa.

Early research on ape genitals revealed remarkable variation in their testicle size and sperm counts. Not surprisingly this relates to their frequency of copulation and thus also relates directly to their social hierarchies. Professor Roger Short from Melbourne University in Australia published pioneering work in this field back in 1977, showing that although a chimpanzee is one quarter the weight of a gorilla, its testes are four times heavier. Female chimps often mate with more than one male during their oestrus (time of peak fertility) whereas gorillas do not, so the chimps have a premium on larger sperm production to ensure mating success. Such seminal observations form a fundamental rule of nature and have since been observed in birds and other mammals, as we will discuss later in the book.

So back to our well-hung little Latino duck. Why would a duck, of all creatures, need to have evolved the longest male genital organ of any backboned creature? The scientist who announced the extended size of the Argentine duck's penis in 2001, Dr Kevin McCracken, did so mainly because the previously longest Argentine duck penis ever measured was only around 8 inches (20 centimeters), so finding another that was almost 7 inches (17.7 centimeters) longer when gravity extended it downwards was indeed an unexpected surprise. The authors of the paper speculated that perhaps it is somewhat like the peacock's tail, whereby males might try to impress females with their elaborate plumage allowing females to assess the better mates. In a similar way, the female Argentine duck might select males with larger penises as part of a display of mating prowess. The sexual selection process identified by Charles Darwin might have driven the males to extreme lengths, so to speak.

Very little is actually known about Argentine duck sex. Can the males put it all the way in without harming the female? Probably not, even though these ducks are reported to be boisterous and promiscuous. The shaft of the Argentine duck penis is in fact covered with spines while only the tip is soft and brush-like. This tip may serve another function, perhaps working with the spines like a bottlebrush to scrape out sperm from a previous male, thus ensuring the mating male with the most appropriately bottlebrush-shaped penis wins the evolutionary competition to inseminate the female and pass on his genes.

All of sexual selection can be summarized simply by the notion of quantity versus quality. The males, having virtually unlimited quantities of sperm, want to inseminate as many females as possible to spread their genes far and wide, whereas the females, with a limited number of eggs in many cases, just want the best quality sperm they can get.

Unlike the external mammalian penis sported by *Homo sapiens*, the duck penis is formed from an extrusion of soft tissue from the inside of its bottom, which comes out when aroused through a combined anal and genital opening called the cloaca (from Latin, meaning 'sewer'). The penis is made erect through filling with lymph fluid rather than blood, which is the usual way to engorge a mammalian penis. Duck penises have another interesting ability: they can literally explode out of the duck's body.

In 2009 Patricia Brennan and colleagues Chris Clark and Richard Prum, all from Yale University, published a paper in a prestigious scientific journal exploring this explosive discovery. Their research concerned the male Muscovy duck (*Cairina moschata*), commonly bred at duck farms for their tasty meat, and the speed of their penis eversion (the coming out or unraveling of the enormous corkscrew-shaped penis). The first step was to study how professional 'duck fluffers' collect semen for artificial insemination. The drakes are aroused by introducing a female duck during the breeding season, and as the male mounts the female his cloacal region begins to swell indicating his readiness to copulate. The duck fluffer then whips the drake off the female and quickly touches the aroused male cloaca whilst holding up a specially made corkscrew-shaped glass jar to catch the penis exploding out of its body and ejaculating as it fully extends. In the lab the researchers measured this phenomenon and found that the entire 8-inch (20-centimeter) duck penis would extend from inside the cloaca to outside and fully erect

Muscovy duck demonstrating rapid 75-mile-per-hour (120-kilometer-per-hour) penis eversion. (John Long)

The Machismo of the Argentine Duck

in just 0.348 seconds. (Next time you are barreling down the highway at 75 miles per hour, just think you are going as fast as a Muscovy duck can erect his penis.)

The Muscovy duck experiment highlights one of the fundamental differences between mammal and bird penises and begs the question: did the mammalian penis evolve separately from that of ducks and other animals? Most zoologists today would argue that it did. More importantly, it also raises the questions of when, where and why the practice of mating by copulation first evolved? If you think about it, the idea that a male of some primitive archaic creature one day decided to put a part of his anatomy inside a rather delicate region of the female, then decided it felt funky enough to ejaculate his sperm, is pretty odd. The evolution of intimate sex through copulation thus poses a lot of interesting scientific questions both behavioral and physiological, and also from the point of view of evolutionary success.

In order to explore these questions we have two main sources of information. Firstly, we have observations from the living world of animals about how and why they mate, and about their relative levels of success, or 'fitness', as biologists like to call that success. Secondly, we have the fossil record of the past life of the planet, which usually tells its stories through scientific interpretation of ancient bones, plants and impressions – graphically, for example, in the two fossilized sharks from Montana dated at around 330 million years old, showing the female biting the large overhanging head spine of the male in readiness for mating.

Had anyone proposed to me ten years ago that I would become an expert on the origins of intimate sex, my reaction would have been one of laughter and denial. Yet my team of colleagues and I have now published a series of papers attesting to the extraordinary discoveries we have made over the past 25 years. These discoveries have revealed not just the origins of sexual intimacy by copulation in our distant ancestors but also the intricate structure of the world's first male vertebrate copulatory organs. The implications of these discoveries in understanding our own evolution are indeed significant, but the most intriguing part of the work is showing in particular how the male sexual organ has evolved through the ages, as revealed through a series of quite unusual fossil finds.

To explain this story we've had to do some truly strange things, like direct the world's first paleo-porn movie and give lucid talks about copulation to museum board members.

The first discovery in this complicated series of finds goes back to 1983 when I was still a paleontology student working on my doctorate at Monash University in Melbourne. I had become fascinated by the ancient armor-plated behemoths called 'placoderms', meaning 'plated skin' and named after the thick bony plates that covered their head and trunk regions. At the time on hand at the university I had some excellent fossil specimens of whole fishes, many represented by parts never before shown in the group of placoderms called 'phyllolepids' – meaning 'leaf scale', as their skin plates were thin and broad, indicating they were a rather flattened fish something like an armored sole. Anatomical features such as jaw parts and the tail region were able to be studied for the first time in this enigmatic group. The material represented a new genus which I named *Austrophyllolepis*. Prior to that discovery only one genus of the group had been known, *Phyllolepis,* and it was mainly represented by northern hemisphere sites, in particular from Scotland, East Greenland, North America, Russia and Europe. Thus the name I chose for the new group, *Austrophyllolepis*, meaning southern *Phyllopis.*

Although I suspected at the time that parts of the *Austrophyllolepis* pelvic fin I was studying could relate to reproduction, I lacked the statistical proof to take it any further (having only a few good specimens). Then, in August 1986, I led an expedition to the now famous Gogo fossil sites in the north of Western Australia, and the first big discoveries that would impact on the origin of vertebrate reproduction were made.

In those days, as an eager 29-year-old just a few years out of university, I was keen to prove myself in the cut-throat world of paleontology. Usually one does that by mounting a major expedition to some remote and often dangerous part of the world in the hope of finding something clearly big and important, along the lines of Howard Carter finding Tutankhamen's tomb or Arthur Evans' discovery of the Minoan civilization on Crete. But when my big chance at fossil renown came along on that at times ill-fated expedition, I would not fully know the secret of one of my discoveries for at least another 20 years.

Gogo Station is near the small inland town of Fitzroy Crossing in the Kimberley region, about four days drive north of Perth. The typical grassy paddocks are surrounded by spectacular jagged limestone ranges, adorned with bottle-shaped boab trees and scraggly bohemias. Now a vast cattle station nearly 60 miles (100 kilometers) across, it was here, 380 million years ago, that myriad bizarre life forms once teemed through an equatorial algal reef. There were masses of primeval fishes, corals, sponges and shells, ancient coiled squid-like animals called goniatites, and schools of bizarre shrimp with peculiar bivalved carapaces – as opposed to today's shrimp which have a series of segments like onion rings. Today the well-preserved remains mainly of fishes and crustaceans at Gogo are encased in rounded limestone concretions (or nodules) littering the valley floors.

The first scientific expeditions to this site were run jointly by the Natural History Museum of London, the Western Australian Museum and Hunterian Museum of Glasgow in 1963 and 1967, then the Australian Museum and the Bureau of Mineral Resources briefly explored there again in the early 1970s.

It took me and my team of volunteers almost a whole week on that first expedition in 1986 before we even started finding anything worth collecting. About one in a thousand nodules contain a fossil fish, and once split open they need to be glued together again for examination back in the laboratory. Our approach was to hit the nodules with a small sledge hammer to see if any fossils were inside them, but soon enough we found that most nodules were barren. Scientists can suffer the same highs and lows of emotion as any self-doubting artist or writer, and at one stage dismal thoughts ran through my brain that the whole trip would be a complete failure because earlier expeditions to the site had cleaned up all the good fossils. Add to that the numerous breakdowns of my hastily purchased off-road vehicle, the two occasions when we were stranded without radio contact, and the time two of us had to walk 10 miles (15 kilometers) through the desert to reach the highway and then get a lift into town for mechanical assistance (I recall we were followed by a starving dingo, who probably thought we would ultimately provide him with some nourishment, at least), and for a while there it wasn't looking good.

After a week or so of hard, back-breaking work in the hot Kimberley sun, however, we eventually stumbled upon areas of Gogo nodules that hadn't been searched thoroughly in the earlier trips. I recall the great excitement of those early days when we began to find superb new specimens and each day added unusual and potentially new species to our haul. By the end of the trip we had bagged a great number of good specimens that included more than a 150 fish and many crustaceans, but most importantly the Gogo fossil fishes we found were very special due to their unique preservation. Most fossil fishes this old (around 380 million years) are two dimensional, crushed flat between sheets of shale. But the Gogo fish skeletons were enclosed in their three-dimensional entirety within the limestone concretion, which would allow us to 'prepare' the bones out of the rock with a weak acid solution. This method slowly and gently dissolves the limestone away, leaving the delicate bones poking out in perfect form, just as the remains of your Friday night snapper dinner might sit on a plate after you've eaten the flesh. As the bones are exposed they are hardened with plastic-based glue and then re-immersed in the acid bath until all the rock is gone and just the bones are left. Then the bony parts are reassembled to make a perfect 3-D skeleton. In some cases, we have the whole fish skeleton embedded within two slabs of acrylic or epoxy resin which, when the rock covering the fish is dissolved away, reveals the delicate bones poking out from an articulated skeleton.

After returning from the field trip I was in heaven preparing the new fossils and writing up descriptions of some of the new forms that I was able to identify.

The most common fossil fishes we found were the remains of placoderms. These fishes ruled the oceans, rivers and lakes of the world for almost 70 million years as the dominant vertebrate life on the planet, yet today almost no-one has heard of them. (Ask a friend or six if they know what a placoderm is and you'll soon get the picture.) Placoderms are generally regarded by most scientists as sitting at the very base of the evolutionary tree of all jawed vertebrates (with humans at the top). As such they are more primitive than sharks and bony fishes, which today comprise the entire living fish fauna, save for a few species of jawless lampreys and hagfishes.

One of these small placoderm fossils was about 5 inches (10 centimetres) long and preserved as two halves in each side of a rounded concretion. At the time of discovery I gave it only a cursory look and labeled the rock with marker pen as 'paleoniscoid', an ancient ray-finned fish (these constitute the vast majority of living fishes, with forms like the salmon, goldfish and marlin, all of which have bony rays supporting all the fins). Back at the lab I examined it more carefully and determined it was in fact an unusual kind of placoderm, a ptyctodontid (pronounced 'tick-toe-don-tid'). I embedded each exposed side of the fish skeleton into a slab of epoxy resin and then prepared the bones out of the rock, ensuring they remained stuck in the plastic resin to keep the fish skeleton in position on both sides.

It seemed to be at the time one of an already described species of ptyctodontid, then known as *Ctenurella gardineri*. Examples of this fish had been found previously at Gogo and were originally studied by my Australian colleague Gavin Young as part of his London-based doctoral work in the mid 1970s, which was eventually formally published with his supervisor, Roger Miles of the Natural History Museum of London, in 1977. So further study of this specimen in detail wasn't a high priority for me. I put it aside and didn't take another look at it until almost ten years later, in the northern summer of 1995.

That year I was fortunate to have four months in Paris as a visiting *professeur* at the Musee Nationale de Histoire Naturelle, situated in the beautiful Jardin des Plantes. I had been set up in the room that had once housed a very famous French paleontologist, the Jesuit priest Father Teilhard de Chardin, one of my childhood heroes. Every day, interesting new finds were revealed within dusty collections in the basement of the old French natural history museum, and wonderful discussions were to be had with other fossil fish experts such as Philippe Janvier, Daniel Goujet and Herve Lelievre. I was riding on an intellectual high of discourse and discovery for most of my time there.

While at the museum I had the good fortune to be able to study some well-preserved ptyctodontid fishes from Germany which were also named *Ctenurella*. These in fact belonged to the very first set of fossils named in the genus *Ctenurella*, and included some beautiful examples showing the skull and jaws intact. In studying these early specimens I soon realized that the original description of them, published back in the early 1960s by Norwegian paleontologist Tor Ørvig, had made a fundamental error in the identification of the restored skull roof bone pattern. This meant that my Australian Gogo specimens, at that point also named as *Ctenurella*, were in fact quite different from the original *Ctenurella* fossils. This was a true revelation for me, as it meant that my little Gogo fossil could now rightfully be referred to as a new genus to make the distinction clear. In writing up a revised detailed description of the German fossils I also presented new data on the previously described Gogo specimens and renamed my ptyctodontid a new genus *Austroptyctodus* (meaning 'southern ptyctodont').

In my paper I described and drew the figures of the specimen known affectionately as WAM 86.9.886 which I had collected in 1986 and prepared embedded in two slabs of resin.

After the paper was published in the French museum's journal *Geodiversitas* in 1997, I turned my mind to other more pressing projects and forgot about it. By then I was working as the curator of vertebrate paleontology at the Western Australian Museum in Perth, so the specimen went on display for the public to see in late 1999 where it remains to this very day in the spectacular 'Diamonds to Dinosaurs' gallery.

Little did I know that this tiny specimen would reassert itself in my life once again another ten years later.

2

The Mother of All Fossils

I have been speculating last night what makes
a man a discoverer of undiscovered things;
and a most perplexing problem it is.

Charles Darwin (1792–1896)
A Century of Family Letters

In late 2004, after 15 years in the public service, I left my position as Curator of Vertebrate Paleontology at the Western Australian Museum to take up the job of Head of Sciences at Melbourne Museum, my hometown. In mid 2005, armed with funds from the Australian Research Council, we ran another successful expedition to the Gogo area and brought back a load of fantastic new fossils to prepare and study.

I recall well the moment on 7 July 2005 when we

found a particularly nice little fish fossil. It was another bright and clear day, with a hot Kimberley sun belting down relentlessly on us. My old mate Lindsay Hatcher was working about 160 feet (50 meters) away from me, scouring the ground intently. A Busselton local with an enduring passion for fossil hunting, Lindsay had been my right-hand man on many fossil-collecting expeditions throughout Western Australia. It was mid afternoon when Lindsay struck a rock with his hammer and suddenly spotted a white patch of bone flashing in the sunlight. He called me over and I looked at it with my hand lens.

'It's just a placoderm,' I said nonchalantly, so he labeled it, wrapped it in newspaper, and bagged it. Round the campfire later that evening I added the date, detailed the locality code and reference number, as we do with every fossil collected at a site. It meant nothing special to us at the time. We could not guess its extraordinary secret until it was prepared out of the rock just over two years later, when its lucky number happened to come up for being treated in the paleontology laboratories of the museum.

That occurred in November 2007 and it would be the first time in over 380 million years that this rock's contents would see daylight. The fossil fish specimen found by Lindsay Hatcher had been mostly prepared out of the rock and it lay as a pile of delicate tiny bones in a white cardboard box. Realizing it was an important specimen, because of its complete head and braincase, I had invited my colleague Dr Kate Trinajstic over from the University of Western Australia to work with me on describing it. Eagerly we took turns looking through the binocular microscope at this strange but beautifully preserved fossil fish, sketching its features, taking measurements of its bones with digital calipers, and always probing for something new in the material.

The unique pattern of skull bones told us that this was a new genus and species. A genus is the term applied to closely related species in any animal or plant group. For example lions, tigers and jaguars are all very similar animals so they belong in the genus *Panthera*, but all represent separate species (*Panthera leo* is the lion, *P. tigris* the tiger and *P. onca* the jaguar). Each species can be defined by its own distinct morphological features, behaviors and genetic characterization in its genes and DNA. In paleontology, as with much of botany and zoology, discovering a new species or genus is always a monumental moment worthy of celebration as we get to extend the known biodiversity of the planet. Plus, it's pretty cool to bestow a new name that will effectively live on in scientific literature forever. That is, if one does a good job of describing it in the first place.

This specimen belonged to the ptyctodontid group. These little fishes had immensely powerful jaws made up of four large crushing tooth plates, two upper and two lower which met like a massive pair of pliers. We guess that most of these fishes fed on hard-shelled clams, snails or other creatures made up of tough outer casings. Our specimen had the skull, jaws and trunk bones completely preserved as well as a section showing the tail vertebrae and fin bones. Less than a dozen of these fishes had been described worldwide from relatively complete remains, so our intact find would shed a lot of new information on the general anatomy and evolutionary position of the group.

I had taken on the preparation of this specimen myself when my preparatory technician, David Pickering, who had begun the job, alerted me to the fact that this fish was a very special one, with many complete bones intact. As we had hundreds of specimens waiting preparation, David was always busy with a batch of Gogo fishes coming out of the acid baths at any one time. I would occasionally take an important specimen and prepare it in my little lab area located in the museum's main building, away from the paleontology labs. This also gave me the opportunity to photograph the specimen at each stage of its preparation to accurately record bone positions.

The acetic acid used to prepare fossils out of rock is common enough; found in vinegar at around 4 per cent, you might have consumed it regularly on your fish and chips. Harry Toombs of the Natural History Museum in London discovered back in the 1950s that bone can be retrieved from limestone rocks when placed in weak solutions of this or formic acid because bone is made of a phosphate mineral, hydroxyl apatite, which is resistant to low levels of these acids. The acid eats away at the carbonate rock but doesn't harm the bone. The preparation process is repeated for about two months before an entire skeleton is finally liberated from the rock. The bones can then be assembled using a reversible acetone-based glue which can be dissolved allowing you to unstick the bones later; it's just like building a model aircraft – quite a satisfying task, as the bones generally fit together perfectly (even when 380 million years old!). Gogo is one the few sites in the world where you can do all this with such old fossils because the rock is just the right kind to dissolve easily.

The Mother of All Fossils

So in November 2007, at around three in the afternoon, we had completed our examination of most of the large bones of the fossil and were left with the job of describing the tail. It was still partly hidden in a small lump of rock, with a series of tiny vertebral elements strung together in neat formation leading to it, because it's better to study it whilst articulated rather than as a jumble of tiny little bits. The tail, even I'll admit, is generally a fairly boring part of the work necessary to complete the paper that would follow from the study, so it was no wonder our minds wandered onto more exciting topics.

Kate was musing over what we should name the fish. I suggested something in honor of Professor Curt Teichert, the famous German geologist who first discovered the Gogo fish site in the 1940s. *Teichertodus*, meaning 'Teichert's tooth', alluding to the powerful tooth plates of the little fish, seemed an appropriate name, and we started labeling all the photos and data with this provisional moniker. (In science, any new species or genus name is not official until the peer-reviewed paper presenting the name is formally published.)

Meanwhile, staring down the microscope at the tail region, I realized I was not able to accurately count all the vertebrae, nor see all the bones associated with the fins.

I suggested to Kate that we risk putting it back in acid one last time, because it was such a promising specimen, but only for an hour or so, and in an extremely weak solution of about 3 to 5 per cent, to try to lift a very thin layer of rock from the bones. After carrying the specimen to the sink in the little zoology lab near my office, I measured out the required dose of acetic acid, mixed it with water and carefully immersed the specimen in a small plastic ice-cream container. Then I poured some of the spent acid water from the previous day's work into the container, to slow down the process and try to avoid damaging any fragile little bones by the bubbling caused in the chemical reaction. It began to fizz very softly, tiny bubbles rising to the surface as the acid ate into the rock.

Whilst the hour ticked away, we joked about how something 'really big' might come out of our study of this particular fossil. With one of the best preserved braincases (neurocraniums) of any ptyctodontid so far found, I was optimistic that we would eventually reveal something exciting about how the placoderm brain was closely related to either that of early sharks or bony fishes, the two major living groups of fishes.

The origin and relationship of the placoderms, an entirely extinct group of fishes that died out 359 million

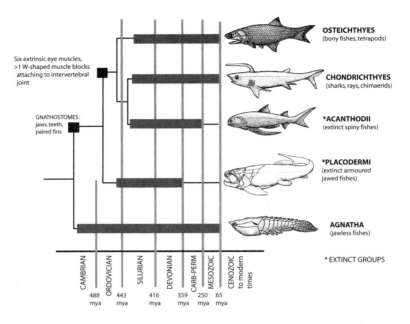

Six extrinsic eye muscles,
>1 W-shaped muscle blocks
attaching to intervertebral
joint

GNATHOSTOMES:
jaws, teeth,
paired fins

OSTEICHTHYES
(bony fishes, tetrapods)

CHONDRICHTHYES
(sharks, rays, chimaerids)

*ACANTHODII
(extinct spiny fishes)

*PLACODERMI
(extinct armoured
jawed fishes)

AGNATHA
(jawless fishes)

* EXTINCT GROUPS

CAMBRIAN | ORDOVICIAN | SILURIAN | DEVONIAN | CARB-PERM | MESOZOIC | CENOZOIC to modern times

488 mya | 443 mya | 416 mya | 359 mya | 250 mya | 65 mya

A simple evolutionary tree of fishes and higher vertebrates. Placoderms are currently thought to be basal (ancestral) to all jawed animals. (John Long)

years ago, was a topic still hotly debated by scientists. Four major hypotheses had emerged. Placoderms were either: (1) closely allied to sharks and their kin (called chondrichthyans, or cartilaginous fishes); (2) closely allied to bony fishes (osteichthyans); (3) the primitive ancestral group to the hypothetical ancestor of both sharks and bony fishes; or (4) an unnatural grouping with some members being closer to sharks, and others below the node where the last common ancestors of sharks and bony fishes diverged. This latter idea implied that all the fishes we called placoderms (an actual class of animals, like birds or mammals) were not in fact a true natural group. We were thinking how this specimen could help us solve the mystery of placoderm affinities if the braincase revealed new anatomical information never seen before. Such new information could help resolve whether the placoderms were more closely related to one or other of the living groups of fishes. We might even score us a paper in one of the top scientific journals, like *Nature* or *Science*.

At the passing of the hour I ducked back into the lab to retrieve the fossil from the weak acid. The newly exposed bones were so fragile that I very gently washed the specimen in slow moving water, so as not to destroy any structures that were newly exposed. Taking it back to my office, I put our specimen straight under the microscope to examine the bones. The tail region close to where it attached to the front of the fish's body was very nicely preserved and I eagerly scanned it to see what had been revealed from the last acid immersion. Most fishes have a pair of fins at the front, called the pectorals, and pair at the back, called the pelvic fins. As expected, our tail showed a section of tiny bones starting from just behind the trunk bones (the chest equivalent) and continuing to the start of the tail fin. But then I spotted something else: a jumble of delicate, almost translucent tiny bones immediately behind the large adult trunk bones, as well as a weird, twisted twine-like structure, like a mineralized rope that wrapped around the tiny bones.

My first thought was that we had evidence of something the ptyctodontid had eaten as its last meal, as clearly another little fish skeleton was here inside it. Then I noticed that there were two sets of jaws, upper and lower, still in articulation as part of the tiny fish skeleton. The jaws were made of four crushing tooth plates, so this indicated that it was another ptyctodontid. That helped me next recognize that the cluster of tiny bones near the jaws were actually little ptyctodontid bones, and they were all of similar shape to those of the big adult fish.

It took about a minute for the penny to drop inside my head. It was one of those sublime eureka! moments that every scientist hopes to experience at least once in their lifetime.

I called out excitedly to Kate, who was working at my desk nearby: 'I think we are going to get that *Nature* paper after all. We have an embryo here inside a mother fish.' It was not just an embryo but undoubtedly the world's oldest fossil vertebrate embryo ever discovered, and clearly one of the best preserved embryos ever found in a fossil!

Kate rushed over to have a look down the microscope. She observed it for a moment or two then agreed with me that we had indeed found a tiny embryo. But playing devil's advocate, she asked me to prove it wasn't an ingested prey item, the big fish's last meal. After pondering this for a short while, we both agreed that the tooth plates were quite specific to this type of placoderm, strong evidence that the little fish embryo was of the same species as the adult fish. This was reinforced by the peculiar shape of some of the isolated bony plates of the head and trunk, almost miniature replicas of the adult fish's skeleton, and so of a quite different shape from plates found in other closely related species of ptyctodontid fishes. In short, the morphology of the tooth plates and head and trunk plates in the embryo more or less closely matched those of the adult mother fish, taking into account some subtle variations expected due to being at different stages of growth.

'But what if it was a cannibal?' Kate suggested.

'Good question,' I replied.

We looked at the position of the tiny embryo skeleton and noted how it was tucked high up near the vertebral column, where the fish's ovaries are meant to be located. The gut or alimentary canal in any fish would be located closer to the lower belly section in the fossil, and our set of tiny bones was not found there.

Kate also noted that the external ornamentation of the delicate little bones was so finely sculptured that this indicated the bones were well preserved, with none of them broken or damaged in any noticeable way. If the little fish had been eaten, we would expect some obvious damage in the food reduction process (crushing by jaws), or gastric etching by harsh digestive acids in the stomach. Neither was evident in our amazing little specimen; I was already starting to call it 'my precious'.

By this point we were absolutely convinced we had a fossil embryo, but we still had no idea what the strange rope-like structure was that had twisted itself around the little fossil. Kate took a small piece of the structure that had come loose in the preparation and popped it in a vial for more detailed examination under the scanning electron microscope when she returned to Perth.

We decided that our find was so spectacular that we should put the specimen back in weak acid just one more time to expose more of the tiny embryo. In the end we repeated our previous procedure twice more, allowing about an hour each time and then carefully checked the specimen until we had exposed the jaws and other embryonic bones enough to fully study them. I also reviewed the photos I had taken of earlier stages of preparation so we could identify the stage when some of the little bones were first being exposed to the air. Some of these had since been freed from the rock and were loose in the residues from previous acid baths. With the photos as a reference I was able to pick these incredibly fragile little bones out of the residues with a wet paintbrush and mount them on a plastic microscope tray-slide.

That night Kate, Heather, my wife, and I cracked a fine bottle of French champagne and toasted the discovery of the mother fish with her remarkable 380-million-year-old embryo. Although it was clearly the most exciting thing I had ever found in any of my 30 years of field expeditions, we also recognized the need to keep it secret, both from the media and our colleagues, so as not to jeopardize its publication in one of the prestigious scientific journals. (Submission guidelines make it clear that such journals are not interested in announcing a major discovery to the world if it has already received media attention.)

A quick scan of the scientific literature had confirmed that, apart from some possible shark-like foetuses found alone (without the remains of a mother) in rocks dated at around 320 million years old from the Bear Gulch Limestones of Montana, USA, the next oldest embryo fossils of any vertebrate were of tiny marine reptiles with long necks called *Keichousaurus* from the Triassic period in China, dated at around 220 million years old. Embryos in the Jurassic-period ichthyosaur (a dolphin-like marine reptile) called *Stenopterygius* and dated at around 160 million years old that were found in Germany had been known for many years, being first documented in 1930 by British paleontologist William Swinton (and discussed further in Chapter 11). Several hundred were known in museum collections scattered around Europe and Britain. Some even showed embryos still inside the mother's body, whilst others were fossilized in the very act of birth – perhaps an emergency abortion of the foetus caused by the trauma of the mother's sudden death.

The foetal fossil sharks from Montana, measuring about one-sixth of an inch (4 millimeters) or so, had earlier been named *Delphydontos* by my good colleague Dr Richard Lund, then based at Adelphi University in New York. They were published in the journal *Science* in 1980 as the oldest known vertebrate foetus fossils, at around 320 million years old. These tiny fossil fish appeared to be newly born or even aborted foetuses, but as none were found with the mother it was never definitely proven that they were evidence of 'live bearing' – they could have been foetuses hatched from an egg. So in our minds we had just pushed back the record for the oldest definite fossil vertebrate embryo by almost 200 million years!

About a week after our discovery I had a phone call from Kate, who had put the tiny sample of the white twisted structure from our fish under the scrutiny of a powerful electron microscope. Cranking it up to several thousand times magnification, she had examined it from every possible angle. She began by saying two words that made me simultaneously smile and go weak at the knees: 'umbilical cord.'

She continued, 'John, it's a fossilized umbilical cord that connects the embryo possibly to the yolk sac.'

'But how can we demonstrate that and convince our peers who will review the paper?' I responded.

'Well,' she said, 'I have identified a number of features that show it has to be a feeding structure. It has pores and vesicles for fluid transfer, and little scars for ligaments that attach to the umbilical structures we see in modern sharks, called "appendiculae". Plus there's an outer epithelium over the porous layer. All up we have four distinctive features that are also found in the modern umbilical cords found in some living sharks.'

I was dumbfounded just thinking about it. Not only had we found a new genus and species belonging to a completely extinct class of animals (placoderms), but we had discovered an amazingly well-preserved and delicate fossilized embryo whose bones were in 3-D format still inside its mother. Now we had the only fossil ever found anywhere showing a maternal feeding structure preserved. It truly nailed it that we had a definite embryo. There was even a vugh (or cavity) in the rock near the rear end of the umbilical cord that was filled with coarse deep yellow calcite crystals. We theorized that this cavity was possibly the relict position for the yolk sac which had deteriorated away, leaving a cavity to be later filled with calcite organic growths.

Yet the most significant part of this discovery had still eluded us and was to come a month or so later when we got together and began writing our paper to announce our discovery to the world. We brought in two close compadres to help us complete the paper: Dr

Gavin Young, a world-renowned expert on placoderm fishes at the Australian National University (ANU) in Canberra and coauthor of the original discovery of the Gogo ptyctodontids published in 1977, and Dr Tim Senden, the scientist who built one of the world's highest resolution micro-CT scanners and a chemist in the Department of Applied Mathematics also at the ANU. They would be instrumental in ensuring our paper covered all bases of the story.

Tim's new technology would enable us to X-ray the specimen in very fine detail and reveal minute parts still embedded in the thin limestone layer that held the tail and embryo together. In late 2007, when I flew up to Canberra to meet with Gavin and Tim. We put the specimen through, working late into the night, and then eagerly pored over the images of the umbilical structure weaving through the rock in 3-D outline. The tiny embryonic plates visible on the surface showed their every feature in clear view, so we could resolve each last detail that had not been visible through the microscope. Now we had everything we needed to start our paper on the world's oldest fossil embryo.

The true significance of our find dawned upon us as we contemplated the discovery over a few cold beers on a hot Canberra afternoon in Gavin Young's backyard. We had stumbled serendipitously upon something much bigger than just the world's oldest embryo. For an embryo to be raised inside the mother, it generally means that the female fishes were not simply laying their eggs in water with the males ejecting sperm over them. They were copulating. They were having intimate and complex sex somewhere near the limey sea floor, around 380 million years ago.

Gavin sipped his beer and mused over the discovery. 'Gentlemen,' he announced solemnly, 'I do believe we have just discovered the first fossil fuck.'

So how the hell were they doing it?

3

The Ptyctodontid Kind of Congress

*An ingenious person should multiply the kinds
of congress after the fashion of the different kinds of
beasts and of birds. For these different kinds of congress,
performed according to the usage of each country, and
the liking of each individual, generate love, friendship,
and respect in the hearts of women.*

Karma Sutra, 1883

To imagine how 380-million-year-old placoderm fish
might have had sex, there is really only one living analog to
turn to for comparison – the class Chondrichthyes, which
includes all living sharks, rays and chimaerid fishes. This
is because our little ptyctodontid placoderms have males

bearing bony extensions to their pelvic fins rather like the pliable claspers of sharks and their kin, used to reproduce by transfering packages of sperm inside the female. So to try to recreate how 380 million-year-old armored fishes might have mated, we first need to understand a bit about mating in living cartilaginous fishes (those whose skeletons are cartilage on the whole, not bone). What follows is a crash-course primer: Shark Sex 101.

If you have ever been close enough to a shark to tell what sex it is then you were probably too close for comfort. But in the safe setting of any of our modern-day aquaria, such as the beautiful Aquarium of the Pacific in Long Beach, Los Angeles, you can observe sharks and rays very close up, sometimes even swimming directly over your head. From this vantage point it's not hard

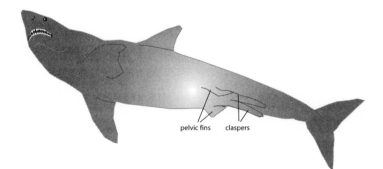

Male shark showing claspers. (John Long)

to tell the sexes apart, as the males have very long lobes extending back from each of the paired pelvic fins. These claspers (also known as 'valva') were once thought to work by grabbing hold of the female during mating. In fact, when they were finally observed in action, they were found to be more than just elaborate grasping structures. They are true 'intromittent' organs, meaning they are inserted deep inside the female to deposit packages of sperm called 'spermatophores'. What they are, in effect, is a pair of penises attached to the pelvic fins.

Sharks and their cartilaginous kin all have this same kind of intimate mechanism, although courtship can be quite variable between the species, and in some it can be downright rough. The anatomy of the male claspers and pelvic region can also differ quite markedly between the many species. Chimaerids, elephant sharks and other members of the holocephalan group, for example, have 'pre-pelvic' claspers (closer to the head than pelvic claspers) as well as the main set on the pelvic fin, and they also bear a small, strange single clasper-like thing on the head of the male called a 'tentaculum'.

Dissecting a clasper reveals that it is a tube supported by a long cartilage rod, this rod being an extension of the internal skeleton of the fish's pelvic fin. As the shark matures, the claspers become more calcified as crystals of calcium carbonate are deposited inside the cartilage of the clasper to make it strong and thus more rigid. In fact this is one of the best ways to gauge maturity in sharks. If, for example, you wanted to determine the age of a male great white shark (*Carcharodon carcharias*), and you happen to be clinically insane, you could swim up to one and cop a feel of his claspers to see how stiff they are; the male of this species isn't considered mature until his claspers are quite hard. Maturity occurs when the male shark is about 10 feet (3.5 meters) long and the clasper, which grows accordingly, has reached a size between 1.3 and 1.5 feet (40 and 45 centimeters) in length.

Claspers act in very much the same way as our mammalian penis does, in that they become erect (from a rubbery, flexible state) once blood is pumped into them. By maturity the claspers are capable of full rotation from their base to face forwards, allowing the fish to mate face to face with its partner. Not a pretty thought, but necessary as out in the open ocean it's difficult to, shall we say, gain purchase. In order to keep them inserted during the sometimes lengthy act of mating, tiny hooks and barbs developed from external scales are often found on the business end of the clasper.

While all known sharks, rays and holocephalan males have claspers, only holocephalans have the accessory sexual organ called the tentaculum on the heads of males. This hooked lobe whose purpose is possibly to assist in the mating process has sometimes been described as a penis on the head of these fishes, this is to use wrong terminology. The little tentaculum might be used to stimulate the female for mating, or to assist in positioning the male fish for mating, but in all honesty we don't know what it's really for – holocephalans seem to keep things private and have never been observed mating in the wild.

Mating in sharks is also something we know very little about on the whole. While some 1000 species of living sharks, rays and holocephalans are now known, my guess from a scan of the scientific literature would be that reproductive behavior is only documented for around 20 to 30 species. Some species have been observed mating in aquaria, like the Port Jackson or horn shark (*Heterodontus*), but eyewitness accounts of sharks and rays mating in the wild are very few and far between.

One excellent account was published in 1985 by Dr Tim Tricas of the University of Hawaii on white-tipped reef sharks (*Triaenodon obesus*). He described the male shark cruising up to a likely looking female and, if he fancies her, keeping close company for a while before biting the female around the neck and front fins. Eventually the male grabs the female by a pectoral fin taking it almost entirely into his mouth. The two fishes rest with their heads on the sea floor (headstand-style), giving them purchase on the substrate whilst the male then begins to insert his erect clasper inside the female. The act is all over within a few minutes for this species. Females of the same species sometimes bear nasty bite marks around the front of the head or on their backs near the dorsal fin, clearly the result of rough foreplay from amorous males.

Interestingly, with mating and courtship rituals frequently quite brutal, the females are often found to be thicker skinned than males, and I'm not talking about their personalities. Actual cross-sections through the bodies of the blue shark (*Prionace glauca*) just behind the pelvic fins show the female's skin layer to be about twice as thick as that of similarly sized males.

In the southern stingray (*Dasyatis americana*), females can mate with more than one male in quick succession (known as polyandry), and the mating act, which involves the female flipping over so the two rays are face to face, takes less than 20 seconds.

General observations of living fishes prove that any fish with claspers present on the pelvic fins is a male, and that all use them for internally fertilizing the female. After copulation takes place, the young are either born alive (called 'viviparity') or are encased with a horny keratinous shell and laid as a few large eggs (called 'oviparity'). Time to apply these few simple facts to our Devonian fishes and see what it all means.

Up until recently only one group of placoderms, the heavily jawed ptyctodontids, had revealed any evidence at all of their reproductive behavior. This story of discovery began when anatomist Professor David Meredith Sears Watson published a paper in 1934 in an Edinburgh journal which showed, for the first time, that certain placoderm fishes exhibited signs of sexual dimorphism (that is, males and females had different body types). Watson had been studying a collection of small ptyctodontid fish fossils from Edderton, Scotland, of a species he called *Rhamphodopsis*. He identified that the males had a long type of pelvic girdle (that is, the internal bones that support the fin) extending back from the pelvic fin, which was relatively small compared with the much larger, heavily scaled pelvic fin in the female. His follow-up study of 1938 also described for more features of these fishes, but in neither work did he recognize that the males actually had the bony equivalents of claspers.

Watson's specimens were fairly compressed, like most Devonian fishes, so not all details of the tiny pelvic fins were clear. It wasn't until 1960 that Norwegian paleontologist Tor Ørvig, at the Swedish Museum of Natural History, described the ptyctodontid fish *Ctenurella* from Germany and he formally identified claspers on the male pelvic fin. Another brilliant English scientist, Dr Roger Miles of the Natural History Museum of London, more accurately described the male and female pelvic anatomy of Watson's fish *Rhamphodopsis* when he published his 1967 paper. Miles was then able to describe a number of smaller accessory bones that differed (from the female) in male ptyctodontids. He identified a single large bony club covered with hooks and spikes as forming the main body of the male clasper. While Miles' summary in that paper was a bit shy and vague, he did lean toward thinking that, like in modern sharks and rays, the claspers must have been used for internal fertilization but did not comment on how the intimate act must have taken place.

In 1963 and 1967, the Natural History Museum, working closely with the Western Australian Museum and Hunterian Museum of Glasgow, collected Devonian fishes from the newly discovered Gogo site in Western Australia. When ptyctodontid fossils were first identified from the fauna, Miles got his bright young Australian postgraduate student Gavin Young from Canberra, to take a closer look at them. Miles and Young published their landmark paper on these fossils in 1977, also making a sweeping revision of the evolution of placoderms at the same time. The most interesting thing about this paper was that it provided the first accurate study of the male claspers from non-crushed specimens of these extinct

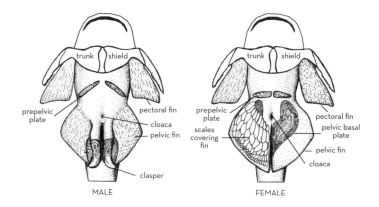

Male ptyctodontid placoderms have bone-covered claspers, whereas females show heavily scaled pelvic fins. (After Miles 1967)

The Ptyctodontid Kind of Congress

fishes. They identified that each male clasper had two sets of external bones covering it, one armed with many sharp spikes that sat on the extreme tip, and a slender curved unit that was plastered onto the outside of the main internal clasper cartilage.

What this meant was that if the claspers were to be internally inserted, there would be problems. Unlike the erectile and flexible cartilage claspers in sharks, there didn't seem to be any easy way for the male to get them in and out of the female. One of my early essays suggested that the clasper might have originally been used to grab the female and hold her cloaca close to the male's so he could shed sperm into her. No convincing evidence has arisen that the claspers of these armored fishes have grooves for transferring sperm as in modern sharks and their kin.

In either case, the behavior required for both courtship and intimacy comes up. Either the male had to get close enough to grab the female and hold her belly to belly using the claspers, or he had to get close enough to push the clasper inside her cloaca. The mechanics and maneuvering required for these actions are mind-boggling.

Our team recently examined all the known 3-D preserved ptyctodontid claspers in museum collections –

there are only a few existing specimens, after all, two in the Natural History Museum in London and one in the Western Australian Museum in Perth. Our conclusion was that these bony structures can be understood as merely covering part of the external surface of the clasper, not wrapping entirely around it. There is still room to move to erect the tissue surrounding the clasper, so this would enable the fish to both stiffen the clasper for insertion and let the organ go flaccid after copulation. A bit of wiggling and to-ing and fro-ing would be necessary to delicately extract the hooked, curved and bone-covered clasper from the female's body, but isn't that half the fun of sex anyhow?

So how did such heavily armored ptyctodontid fishes have sexual intercourse? Firstly, like modern small sharks and rays, they must have rested on the substrate, the sandy sea floor, to enable the act to proceed. We assume that courtship behavior of some kind occurred in order for the male to 'prepare' the female for mating – what biologists often term the 'release phase' and what we others call 'foreplay'. With no direct evidence of how this might have occurred, we can only look to the closest living analogues, sharks, and suggest that they swam close to each other as the male began to nibble on the female's fins or tail to initiate the necessary flow of hormones needed for mating to proceed. Once the female was ready, she might have laid on her back exposing her cloaca to the male, or simply embedded her head nose-down into the sandy seabed. In that position the male would have required a stronger grip, so perhaps he used his trunk plates to hook up, so to speak, with those of the female armor. By placing the belly surfaces close to each other, the male might have been able to maneuver the female's body into the right position for mating using their flexible pectoral fins.

By this stage, the much-stimulated male would have been pumping blood into his claspers to make them erect. By gripping the female's trunk armor or pectoral fin, he could maneuver his erect clasper carefully towards the female's open cloaca. Perhaps hormones had dilated the female's cloaca for ease of mating – anything to help deal with those darned hooks and spikes would have been a blessing. The hooked clasper would need be inserted with much sideways body motion from the male. Once inside her, the curved, bony hooks and ridges would no doubt hold the clasper in place whilst the male transferred his sperm.

Whether this transferral was a package of sperm

(spermatophore), as many sharks pass today, or just an emission of sperm in fluid is not known, nor will it ever will be, I imagine. Whatever the delivery method, it did the trick as there were many ptyctodontid fishes to be found in the seas during this Devonian period of Earth's history. When the act was over, the deflated clasper could be removed by the male thrashing from side to side, no doubt assisted by the female wriggling. The male would probably then swim away, his mission accomplished. We have no way of knowing if they ever stayed in touch.

The female on the other hand now had a new responsibility to take on – the caring of just-fertilized young as they grew inside her body. With internal fertilization comes the joy of pregnancy, which in biological terms involves a major internal physiological rewiring job like never before. Where other fish simply leave the fertilized eggs to hatch by themselves, the pregnant mother fish must now nourish and care of both her young and herself simultaneously. There are many more risks for a mother carrying around large unborn young inside her compared with fishes that abandon their young after eggs are laid. If the mother is attacked and eaten by predators, for example, then the whole point of the mating exercise is lost. Perhaps at this stage of evolution male parental care might have kicked in, with the father helping to guard the pregnant female. Or, more likely, the pregnant female simply hid away in a safe spot whilst her developing young grew bigger each day, feeding from their yolk sacs inside her. She would still be able to feed on the abundant clams and snails living on the reef as long she was constantly wary of predators and kept close to her secure little cave or nook.

It was at this stage of the pregnancy that *our* mother fish, bearing just one large unborn foetus inside her, met an untimely fate. No attack marks are evident on her bones, nor are there any physical signs of a violent struggle, as might exist in the form of wounds or pathologies. Perhaps it was something simple that did her in, like venturing into a zone of deoxygenated water, killing her quietly and unexpectedly. Her body would have then floated to the surface and eventually, after some initial decay, sunk down to rest in the gloomy grey muds accumulating in the stagnant inter-reef basin. Her body soon became fossilized and would, some 380 million years on, be exposed on the surface of the Earth at a remote place called Paddys Valley, about 60 miles (100 kilometers) from the outback town of Fitzroy Crossing.

4

Announcing Fossil Sex to the Queen

Scientists in film and television tend to be depicted as villains, geeks, or jerks. Rare indeed is a Hollywood script or scripted drama that tells a story about science that's both serious and entertaining. That strongly affects how we think.

C. Mooney & S. Kirshenbaum, 2009
Unscientific America

After making a big discovery many scientists immediately descend into a period of despair and worry about whether or not it will be accepted by one of the leading science journals. Even with momentous finds, it can take just one doubtful reviewer expressing even a modicum of criticism for a paper to be rejected by the editors. Most scientists dream about being accepted by journals like *Nature* or *Science*, as the news will then be taken seriously by journalists around the globe, ensuring the story gets widespread media publicity. More importantly, the people who review grants for research funding are often keen readers of such journals, while the politicians who decide on science funding are often readers of the general media. So the sad fact is that a prestigious publication combined with massive media coverage undoubtedly increases the probability of continuing or future funding opportunities for research.

Since our lab discovery in late November 2007, we had been working non-stop to craft a paper for publication. The first step had been examining the specimen through the radiographic eyes of the micro-CT scanner in Tim Senden's laboratory in Canberra. These 3-D scans enabled us to produce new computer-generated images of the fish's bone features. I was also working on some computer graphic artwork showing its anatomical features clearly identified and labeled. The aim of such a paper is to make its key message crystal clear, even for non-experts in the field of paleontology, so there can be no doubt why we were proposing that this discovery was a major breakthrough in science.

After much editing and re-editing, we finally completed our first draft to the point where all the co-authors were happy with it. Then one night after work in early January 2008, I uploaded the file to the author-submission site of *Nature*. A few days later we received a standard reply saying that it would be sent out to review. Hooray! We had passed the first big hurdle: the editors were interested enough in our paper to have it assessed.

The peer-review process is the most worrying phase of the whole submission process. Our paper was sent to three reviewers, each of whom has to confirm to the editor that it is indeed a significant contribution worthy of publication in the journal. They may point out problems that can be easily corrected in the paper, such as an error of fact, a typo or even a misquoted reference, but if any of them has a major criticism of the work, such as a disagreement with the interpretation of the primary data, then the paper will probably be rejected. Two anxious weeks later, the reviews for our paper came back with a message from the editor that while the paper could not be accepted in its current form, if we addressed the various points raised by the reviewers they would reconsider it for publication.

About a week after that, as I was dozing in bed one Saturday morning, a major revelation came to me out of nowhere. I had been mulling over the discovery of our mother fish, when I suddenly remembered *exactly* where I had seen similar tiny clusters of bones in another Gogo specimen. Leaping out of bed, I ran naked to the computer in my study where thousands of images of all the fossil fishes I had ever studied or written papers on were filed away. A few clicks later and I was staring at a most remarkable Gogo specimen – the articulated ptyctodontid that I had collected in 1986, prepared in 1987, studied in 1995 and finally published in the French journal *Geodiversitas* in 1997. I had named it *Austroptyctodus* and identified the clusters of scales along the body as a rather odd feature at the time. No other articulated ptyctodontid described from the famous sites in Germany, Scotland or North America had such scales, although as I had also found another Gogo ptyctodontid with scales, *Campbellodus*, it seemed a reasonable enough description when I wrote it.

That morning, enlarging my color photos of the *Austroptyctodus* specimen WAM 98.9.668, I realized that these clusters of scales were actually tiny embryonic bones. Each cluster was a complete tiny skeleton, and some showed traces of a mineralized umbilical cord near them. I had just identified the second known placoderm 'mother fish'. But this little female was indeed very special: she had died carrying triplets. On public display in the Western Australian Museum's 'Diamonds to Dinosaurs' gallery for the last eight years, not one person, not even the many visiting expert fish paleontologists who came from far and wide to pay homage and study the Gogo fish collection, had noticed her delicate condition.

The timing of this new discovery was perfect, given we were in the process of revising our paper for *Nature*. I immediately sent an email to the journal's biological sciences editor, Dr Henry Gee, informing him that we had found a second specimen, this one with three embryos intact. He advised us to include it in the revision of the figures, responding to one of the reviewers' requests to show more details of the find through the photos. Adding a color photo of the new *Austroptyctodus* specimen with its multiple embryos, our case for placoderm viviparity (live birth) looked solid. We sent the revised, updated version of the paper back in early April and waited for the verdict.

We were in luck. This time around the reviewers all fundamentally agreed with our interpretation that the fossil represented an unborn embryo with a mineralized umbilical structure preserved, as well as accepting the additional evidence provided by the newly recognized *Austroptyctodus* specimen. Our paper was finally accepted by mid April, and we were relieved and excited, as few of us had previously published in *Nature*. In my 28 years of research, I had only had two previous papers in the journal. This time we expected news of the world's oldest mothers, and of the origins of intimate sex, to attract a good deal of worldwide media interest.

Not long afterwards, I received a phone call from Susannah Elliott of the Australian Science Media Centre. I was suspicious at first; we had kept our discovery tightly under wraps while waiting for *Nature* to make a decision on publication of our find. Susannah told me that next month the Royal Institution of Great Britain was holding a ceremony in London to celebrate the completion of its 22-million-pound renovations to the grand old building. At the same time they would announce the opening of an Australian branch of the Royal Institution in Adelaide. The gala event would be attended by a host of UK science celebrities, such as Baroness Susan Greenfield and Sir David

Attenborough. Assorted members of the Royal Family, including Queen Elizabeth herself, would also be there.

The Australian Science Media Centre was looking for a high-profile story to promote as part of the celebrations, and it turned out that *Nature* magazine had been in touch with them to discuss the possibility of using *our* paper as the headliner. Nothing had yet been decided, but if *Nature* chose our paper then news of our discovery would be unveiled to the world from a special dinner held in Adelaide. A live satellite link-up to London would mean we could make the announcement simultaneously to large gatherings of Australian and international scientists and media, as well as all the UK dignitaries and attendants.

Now we were more excited than ever – how many scientific papers are granted such a splendid launch? After much impatient waiting, *Nature* magazine confirmed that our paper was the one chosen, so long as we were able to attend the satellite hook-up in Adelaide on the night of 28 May (the day our paper would be officially published in the journal). There were some initial reservations by my media team at Melbourne Museum that I might not make it back from the event in time for our official press conference at 10 am on the morning of 29 May, but I assured them I'd be on the very first flight home on the morning after the dinner and at the museum well before 9 am.

The next task was to figure out how best to display our 'precious'. We realized early on that the actual fossil of the fish and embryo was very small and rather unspectacular to a look at for the lay person. While a massive dinosaur skull might attract media attention, our little mother would need some help. My boss, Dr Robin Hirst, came up with the idea of making an animated movie showing how the fossil may have looked in life, and I was charged with overseeing the production of a 30-second movie clip for the world's media. This was a lot harder than it sounded.

Firstly we needed a life-sized model of how the pregnant mother fish might have appeared, and one of the museum's skilled model makers was set the task. After being checked for accuracy, the model was sent off to be 3-D laser scanned to create files the computer animator could work with. After several sessions discussing movement, swimming styles, the reef environment and possible modes of birthing, the final 30 seconds of animation were completed. The clip shows the fossil, complete with embryo and umbilical cord, and then brings the fossil to life as a living, swimming fish which then gives birth to its pup. (You can see it yourself on many websites, including YouTube; search for *Animal Armageddon* series, the mother fish episode.)

On the afternoon of 28 May, Melbourne Museum staff began setting up our temporary display, so that immediately following the next day's press conference the delicate little fossil of the mother fish could be viewed by all in the museum's foyer. Next to it was the life-size model of the pregnant fish as it may have looked in reality, and nearby a large screen would project our animated clip of the fossil coming to life and giving birth.

After the display was ready, I headed off to Adelaide. With no time to dawdle, I dropped my bag at the hotel then grabbed a cab to the venue to meet up with Susannah from the Australian Science Media Centre and two of my co-authors, Kate and Tim. Despite about 80 Australian leaders in science and media expected to attend the event, the announcement of our discovery was a secret until the media embargo was lifted later that evening.

While the guests were still arriving, host Robyn Williams, guru of Australian science, hived me off to a nearby couch to do a sneaky interview for his national radio program, *The Science Show*. Obviously he'd been let in on the secret! When he nabbed Kate and Tim for some quick sound bites, I wandered back to the party. The stage was arranged with a table for us three authors and a large screen for projecting the UK link-up, while many beautifully set dining tables were scattered throughout the room. The air buzzed with the chatter of an expectant and knowledgeable crowd that included leading Australian university deans and professors, the newly appointed Australian head of the Royal Institution, and astronaut Andy Thomas, amongst others.

After a quick dinner and nerve-settling glass or two of a fine Barossa Valley red, it was showtime. Kate, Tim and I headed up to the stage while Robyn took to the microphone to introduce the evening's schedule of events. Magically, the white screen behind us sprang to life, beaming in the crowds from the Royal Institution in London. After some introductions and speeches on both sides congratulating the Royal Institution on its newly renovated buildings, Sir David Attenborough entered the conversation. We had named the mother fish *Materpiscis attenboroughi*, meaning 'Attenborough's mother fish' in honour of the great man – Sir David was the first person to present the significance of the Gogo fossil sites to the world in his 1979 television series *Life on Earth* – and Robyn Williams asked him how he felt having the new fossil named in his honour. His reaction on the big screen was one of delight. (We had also nicknamed the fish 'Josie' after my mother, and she was equally thrilled with the honour.) Finally it was our moment. Standing to walk to the microphone, it felt like a whole school of tiny placoderms had taken up residence in my belly. But I somehow managed to announce to the world, and the Queen, that we had found the world's oldest fossil embryos, one still intact and complete with umbilical cord. This meant ancient fishes were having a complex kind of copulatory sex about 375 million years ago for 'fun' – the assumption being that animals before these placoderms had been having sex, but not in such an intimate physical manner. Six quick slides and our animated video clip later and my presentation was done, then Kate stood to explain her work on the preservation of the umbilical tissues in our remarkable fossil, followed by Tim's presentation on ultrafine scanning techniques.

Questions were then invited from the waiting British media contingent. Immediately I was seized upon by eager tabloid journalists to elaborate on my statement that sex was 'fun' for fossils fishes. I explained that this was the first known case for fishes, our distant ancestors, that involved the male copulating with the female rather than spawning in water like almost all fishes today do. So it must have been 'fun' (that is, compelling) in the strictly biological sense or they wouldn't have evolved such a hopelessly complicated way to do it.

Many follow-up questions later the camera turned to the Royal Family members quietly watching in the background, and Robyn Williams grabbed the opportunity to ask the Queen if she had any questions about the fossil discovery. She didn't, but her husband, the Duke of Edinburgh, did pipe up and ask us politely what the fossil fish might have looked like (perhaps he hadn't seen the animation).

Around midnight, Team Mother Fish sat down to share one last congratulatory glass of champagne together. The next day the story would break to the world's waiting media, and we all had to catch ridiculously early flights home in order to attend press conferences in our respective states. Thoroughly hyped up on adrenalin, I managed only about two hours sleep before getting up and heading to the airport in the pitch dark to fly back to Melbourne.

At Melbourne Museum I gave the same brief presentation to a room packed with local and national media, and that night the story of the mother fish featured on nearly every Australian television channel. Soon media interest from around the world began to filter in, in particular from France, Germany, the United States and Spain. That day, and for the next week, my co-authors and I were flat out giving radio and phone interviews to the world's media. Saturation was higher than for any other science story I had ever been involved in, including the famous Nullarbor Caves fossil marsupial lion, which was front-page news in Australian newspapers in July 2002. Clearly there was an avid audience for news on ancient intercourse.

How does one gauge the success of a good science media story? One method is by counting the pages of media coverage costed at advertising rates, and the minutes of television and radio airtime at similar commercial rates. These figures can be then adjusted according to set formulae that PR professionals use to evaluate such stories. Our museum PR team did such an evaluation on the mother fish media and estimated that it generated around two million dollars worth of media coverage.

In July it featured on the cover of *Australasian Science*, and later that year the magazine awarded us the

2008 Australasian Science Prize in recognition of the significance of the discovery. At the end of 2008 US science magazine *Discover* published its annual list of the hundred most significant discoveries across all facets of science; the mother fish featured in the list as one of only three paleontological stories. The discovery of the world's oldest live birth even made it into the 2010 edition of *The Guinness Book of World Records*, with a photo of yours truly holding up a model of the fish, and was the cover story for the January 2011 issue of *Scientific American*.

Another way to gauge media and community saturation with any new paleontological or biological discovery involves the creation of a new scientific name. I call this the 'Google factor'. The day before the mother fish's official scientific name, *Materpiscis*, was published in *Nature*, a Google search on it would have resulted in no website hits, because the name was non-existent as far as the internet knew. Within a week of its publication, a Google search on *Materpiscis* revealed that it was mentioned on close to 50,000 websites around the world. Even in 2012, a search on *Materpiscis* will yield close to 52,000 hits.

5

Paleozoic Paternity Problems

Sexual intercourse began
In nineteen sixty-three
(which was rather late for me) —
Between the end of the Chatterley ban
And the Beatles' first LP.

Philip Larkin
'Annus Mirabilis'

Immediately after our find was published, we began to suspect that more examples of fossil placoderm embryos could be lurking in the forgotten drawers of other museum collections. Our mother fish had inspired a clear paradigm shift in thinking about early reproduction in ancient backboned animals. We knew that some of these ancient fossil fishes, the enigmatic placoderms, had copulated to produce live young, and with the detailed study we had done we had the jump on our fellow scientists and knew that there must be others to be found. Would they all belong to the strange little ptyctodontid group, or had other kinds of early primitive fish also developed similarly complex ways of mating with each other?

A colleague, Dr Zerina Johanson, who worked with us on placoderms when she was based at the Australian Museum in Sydney, had recently moved to the Natural History Museum in London with her paleontologist husband, Dr Greg Edgecombe. For a museum to take on a paleontologist husband–wife team and give them both permanent positions was undoubtedly rare, but both were brilliant scientists so any major museum would have jumped at the chance to have them on staff. It was there, under the expert eye of Zerina, that the next big clue in our story was identified, only a month or two after our announcement of the mother fish. The merest hint from Zerina of what this discovery could mean for science was exciting enough to entice both Kate and I to London as soon as possible.

Back in the English summer of 1982, when I was 25

years of age, I made my first trip to the Natural History Museum in London. It was also my very first trip out of Australia. Funded largely through the sale of my much-loved Honda 750cc motorcycle, the planned two-month trip to England and Europe would enable me to attend an international conference on paleontology and vertebrate anatomy at Cambridge University, and to study fossil fish collections held in the eminent museums of England, Scotland, Sweden, Denmark, Norway, France, Italy and Germany. Museum studies like these were necessary to make comparisons between well-known fossil fish specimens and new fossil species I had discovered in the course of my thesis work.

In that month in London, set up in a spare office in the paleontology department, I saw little more than the inside of the Natural History Museum. When friends back in Australia quizzed me about the Tower of London, Buckingham Palace, Kew Gardens or Westminster Cathedral, I sheepishly confessed that I had missed them all – but I did see some amazing fossils and hung out with some living legends of fossil fish paleontology! Not everyone appreciated my priorities.

While maybe not the most culturally enlightening of tours, that first trip to London proved invaluable in sorting out the finer points of fish anatomy needed to complete my PhD dissertation the following year. It also established a professional network of colleagues, including Dr Roger Miles, the doyen of Devonian fishes, with whom I could correspond about all things paleontological in the years to come while working in isolation in Perth. And on subsequent return trips to the Natural History Museum, I've always felt quite at home back in 'my' office.

Arriving at the Natural History Museum in London on another of these trips in August 2008, just a couple of short months after unleashing our mother fish on the world, I was warmly greeted by the new curator of fossil fishes, Dr Zerina Johanson. There were still a few of the old brigade from my 1982 museum stint around, though Dr Miles had long since retired. Zerina set me up in my old room and I began at once to rummage through the museum's drawers of specimens lined up in aisles according to classification. For anyone familiar with the evolution of fishes and how the hierarchy of their classification is ordered, it was quite easy to find any particular specimen.

I had flown halfway around the world on this particular mission to examine two specimens – fishes of the genus *Incisoscutum* – which were of very special interest to us. One of the most common small armored placoderms found at Gogo, there were many good *Incisoscutum* specimens to refer to when Kim Dennis-Bryan and Roger Miles published their detailed description of the little fish back in 1981. Two unusual specimens they discussed contained the remains of what they described as 'undigested stomach contents', comprising the fragile bones of smaller placoderm fishes preserved inside the gut region. We had a hunch they were something altogether different.

Zerina had picked up on this immediately after the publication of our paper on the embryo in the mother fish while she was working closely with Kate Trinajstic and me on other fossil fish projects. Discussing whether or not these were really digested food remains or could indeed be more fossil embryos, we exchanged photos and excited emails for weeks. Since first seeing the images I had developed a burning, almost addictive desire to examine the potential embryos in person so we could be absolutely sure about the discovery. If we were correct, this discovery would be of much greater significance than the mother fish with her unborn embryo: it would be truly profound news throughout the scientific world.

The reason for this was simple, but requires a little background knowledge. Our mother fish, *Materpiscis*, which we named after David Attenborough, belonged to the particular group of placoderms called ptyctodontids, in which clear evidence of sexual dimorphism, as we have noted, had already been demonstrated: males had claspers and females didn't, exactly as we see in sharks and rays today. This sexual dimorphism, as we have noted, was first identified in ptyctodontids back in the 1930s by the famous British paleontologist David Watson, but it wasn't until the 1960s that Tor Ørvig, at the Swedish Museum of Natural History in Stockholm, correctly identified claspers on the pelvic fins of the male ptyctodontid fish *Ctenurella*.

Our fish *Incisoscutum* belonged to a completely different group of fishes, in fact the largest known group of placoderms, the Arthrodira (meaning 'jointed necks'). This diverse group comprised the vast majority of known placoderm species (more than 250) and had until this point never shown any observable evidence of sexual dimorphism, despite thousands of well-preserved specimens in many fine museums around the world. The arthrodires were not even *suspected* by any scientists of having reproduced by intimate acts of copulation. It had been long assumed that they simply spawned in water, like most fishes do today, based on the understanding that they didn't have claspers like ptyctodontids and combined with the evidence of fossil deposits where very tiny individuals had been found. These included forms like *Groendlandaspis*, which I had visited South Africa to study in 1994 and 1996, and had written a paper on with my South African colleagues. These finds of complete tiny armors – the skeletons that enclose the fish completely around the head and trunk in thick bony plates – of arthrodires only an inch or so long hinted more of them being hatchlings from a nursery site rather than larger live-born young.

The 2008 trip to London confirmed what we had expected: the small collections of tiny plates found in the two *Incisoscutum* specimens were both unborn embryos inside mothers. The delicate little plates were undamaged, not etched by gastric juices, and each tiny plate bore a very delicate ornamentation on its external surface, a well-known feature reported in other juvenile specimens of many placoderm fishes. Furthermore, by turning over the actual specimens I could examine them in ways previously unphotographed, following the outlines and features of the juvenile bones in detail. But while we searched high and low for traces of mineralized umbilical cord structures, we could not identify these in these specimens.

Nonetheless, the implications of the discovery were clear: we had uncovered a new and startling fact about the sexual habits of these ancient primitive fishes. In the face of almost two centuries of detailed study, it took two specimens randomly collected from the far north of Western Australia in the 1960s to reveal the most intimate secrets of the largest group of placoderm fishes, the arthrodires. We now knew that they had copulated in the shallow warm equatorial seas of northern Gondwana – as Gogo was at the time the fossils were laid down – and then raised their young inside the mother until they were large enough to be born and fend for themselves. Hardly primitive behavior at all, and certainly not expected in this group that hadn't shown any signs of special mating organs on any of the previously found fossils.

And so the question of Paleozoic paternity arose. While we had identified females with embryos in them, on closely studying their pelvic fins we found no evidence of anything unusual that would indicate some kind of specialized genital apparatus on a male to fertilize the female with. The males must have been out there somewhere in the fossil collections of the world. We had found the mother fishes, so where were the daddies?

Over the next few days we followed up our discovery, looking for further evidence in other arthrodire specimens from Gogo and other classic fossil sites. But there was nothing – we could find no evidence at all of males bearing clasper-like structures in any of the museum's fine Gogo placoderms. So next I examined all the known specimens of the common arthrodire *Coccosteus* from the Old Red Sandstone deposits of Scotland in the Natural History Museum collections (no mean feat at several hundred fishes) to see if I could identify embryos or any special features in the pelvic fin offering clues to how they might have mated. *Coccosteus* was similar in many ways to *Incisoscutum*, but again, no result. How did these ancient males to the deed?

The evidence we sought would soon rear its head, so to speak, right under our very noses, and from within our own collections back in Australia.

As in the case of the unidentified triplets in the forgotten specimen I had collected in 1986, it was another old placoderm specimen that I had worked on in 1984 that would turn out to be crucial to understanding how mating structures first evolved in backboned animals.

In February 1983 Australia hosted its first international gathering of fossil fish experts from around the world in Canberra, bringing together fossil fish aficionados from as far away as China, Estonia, France, the UK and the USA. At the time some of the most exciting new discoveries in the early fish world were coming out of China, so this opportunity to meet colleagues and hear news of their latest finds was fantastic to me, a lowly doctorate student at the beginning of my third and final thesis year.

I was presenting a paper at the conference about the bizarre flattened placoderm fishes called phyllolepids that I had been studying from the Mt Howitt site in central Victoria. Up until the 1960s only one articulated and complete specimen of a phyllolepid fossil had ever been found, dating from the Late Devonian of Scotland. The new specimens uncovered at Mt Howitt in the sixties were not only complete fishes with the tail preserved for the first time, but also showed the first fine detail of the jaw, ear–stone (otolith) and other features that were poorly known or not previously recorded in placoderms in general.

I first visited the Mt Howitt fossil site in 1980 when I started my honors thesis on the geology and paleontology of the region. The fish are found in laminated shale outcrops along a dirt road cutting near the base of Mt Howitt, close to the sparkling, trout-filled waters of the Howqua River. Apart from being one of the most stunning localities I've ever had the privilege of working in, it also happens to be one of the few sites in Australia where complete articulated Middle Devonian fishes representing individuals in all stages of growth are known to occur. The site was discovered by geologist Mark Marsden in the 1960s whilst mapping the region that today spans the eastern edge of the Great Alpine National Park in the highlands of eastern Victoria. Then in the early seventies, Professor James Warren and his team from Monash University excavated the site over several field seasons, carefully removing layers of black shale and splitting them to find the fish fossils. All were taken back to Monash University to await any unsuspecting student who might think to study them for a thesis.

In 1980 I took on one of these groups of Mt Howitt fishes for my honors thesis, a ubiquitous little placoderm known as *Bothriolepis*. On completing my honors degree, I launched straight into a PhD thesis looking at the Mt Howitt ray-finned fishes (palaeoniscoids). Along the way

I sometimes became sidetracked with the other groups of Mt Howitt fishes, and it didn't take too long before the fascinating, flat phyllolepids took my fancy.

Preparing the Mt Howitt specimens was almost the opposite of how we later prepared Gogo fishes. Unlike the Gogo fossils, the deeply weathered bone of the Mt Howitt fishes wasn't well preserved. The standard way to study the specimens was to immerse both halves of the split fish in the rock (known as the part and counterpart) overnight in a weak hydrochloric acid solution, then gently scrub the bone away from the black shale on which the fish was imprinted. The cleaned-out slabs of shale were washed in water, and the surfaces that held the preserved original shape of the bones were ready to be cast in latex rubber. When the bones were represented on the latex rubber peel, the surface could be photographed to show up the finest details. While nothing of the original fossil bone remains after this process is completed, we have an excellent record in the rubber latexes of what the bones looked like.

The phyllolepids had been an enigma for many years. Scientists at first thought these fishes were strange jawless forms related to today's lampreys, though the Mt Howitt finds would eventually demonstrate that they had jaws with teeth. In 1982 I gave a brief overview of the Mt Howitt fishes to a paleontological meeting at Cambridge University, England, noting that as the phyllolepids from this locality showed no evidence of notches for the eyes, my guess was that they were blind.

At the end of my talk, an old and well-known British paleontologist named Stanley Westoll, who was sitting in the front row, came up and told me that he agreed with the blind phyllolepid idea. A few weeks later I took up an offer to visit Stanley and his vast collection of placoderm casts and specimens at his house in Newcastle. As he and I mused over the mysterious phyllolepid group, I resolved that I would look further into them. The upcoming gathering of early vertebrate paleontologists in Canberra in February 1983 looked to be my opportunity to clarify a few issues regarding the mystery of the phyllolepids.

My presentation at the 1983 meeting on the Mt Howitt placoderms went very well, and my colleagues urged me to write up my findings for a paper to be published in the conference proceedings. I had determined from their head plate shapes that the fishes represented a new genus and, as mentioned earlier, I named the genus *Austrophyllolepis*, meaning 'southern

Phyllolepis' (at the time only one other genus was known, *Phyllolepis*, from the Late Devonian of East Greenland, Europe and North America).

Although most of the descriptive work on the new fossils had been routine, I was troubled over one small aspect of them: the strange bones in the region of their pelvic fins which I had never seen the likes of in any other placoderm fossil. The pelvic fin consisted of a flat, broad plate with a peculiar long tubular bone directed to the rear end of the fish. I identified this bone as a 'metapterygium', which at the time was seen to be a common component of placoderm fin structure, so nothing unusual. I had an inkling of an idea that, as only some specimens had these bones, they might even be sexually dimorphic features, with the long trailing pelvic fins an equivalent of claspers on male fishes. Without enough hard data (no pun intended) to be statistically valid, in the end I left the conclusion open, stating that I suspected these structures could be used in reproduction but lacked the evidence to show this conclusively.

Twenty-five years later, in our 2008 hunt for the world's oldest vertebrate willy, we decided to examine the pelvic fins of all the well-preserved placoderms we knew of. Kate and Zerina would re-examine the Gogo specimens in London's Natural History Museum and the Western Australian Museum respectively. From my base back at the Melbourne Museum, where all my earlier doctorate material was housed, I was to pull out the best specimens showing pelvic girdles, make new latex peels of their structures, and try to find any details previously overlooked in my earlier study.

Revisiting those fossil specimens decades later was an emotional experience for me. Firstly, there's the flood of data as you remember each and every specimen, especially the special ones found in the field, or others that were difficult to prepare or interpret. But these specimens also brought back vivid memories of old friends, family and the birth of my first child which occurred as I struggled to finish the manuscript describing the fossils – the real intangible value of fossils like this can only be known to their discoverers and describers. But re-examining these same fossils again in 2008 also meant that their true nature was finally revealed, not by some mysterious intuitive process, but simply by virtue of the new science that had been published in recent years about fin and pelvic girdle development in primitive fishes. The strange tubular bone pointing away from the fish's head that I'd labeled as a 'metapterygium' I could now see bore a closer resemblance to a similar elongated bone called the 'basipterygium' found in the pelvic girdle of all sharks, rays and holocephalans.

The basipterygium would turn out to be our key to understanding the origins of vertebrate copulation. By comparing the elongated basipterygium of the Mt Howitt *Austrophyllolepis* specimen with those seen in the pelvic fin bones of today's sharks, rays and holocephalans, and in many groups of primitive bony fishes, we saw an obvious similarity that was previously not picked up by any placoderm researcher: only sharks, rays and holocephalans, fishes that mated using copulation, possessed a long basipterygium bone in the pelvic fin. This was strong supporting evidence, along with the newly found embryos in the Natural History Museum's *Incisoscutum* arthrodires from Gogo, that these arthrodires, the most widespread and diverse group of placoderms, must have mated using copulation, as did the ptyctodontids. Furthermore, as the

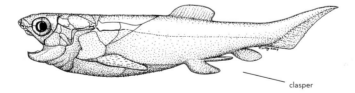

Restoration of *Incisoscutum*, a small arthrodire placoderm fish from the Gogo

phyllolepid placoderms were firmly situated near the base of the arthrodire evolutionary tree, it was likely that all other arthrodires mated using copulation as well.

One other peculiar fact was uncovered through my restudy of the phyllolepid pelvic bones. The robust and elongated basipterygium did not taper to a smooth ending but stopped abruptly in a sharp 'articulation facet': this was clearly not the end of the fin – something else had to be joined to it. Maybe, we thought, this was where the clasper attached onto the fin.

By late 2008 we had still not yet found our smoking gun – the male claspers – in any of the Gogo arthrodires, but we did find some exceptional specimens showing a basipterygium projecting from the pelvic fin. Indeed, on nearly all the known Gogo arthrodires that had the pelvic girdle preserved, the articulation facet for the large backwards-facing basipterygium was always there. The actual basipterygia we found were all in specimens of *Incisoscutum*, the same fish in which we had earlier identified the new embryos at the Natural History Museum in London.

The Gogo specimens showed the basipterygium in clear 3-D shape. The end of the bone did not taper into a point but, exactly as in our phyllolepids, and had a facet for the articulation of another pelvic fin element. So by looking at sharks and rays where the basipterygium supported junction cartilages and then the clasper in males or a shorter cartilage element in females, the evidence was clear to us that we had a similar situation in placoderms. We now had a good case to write a second paper showing that arthrodires had intimate sexual reproduction, and some at least gave birth to live young.

By the end of that year Kate, Zerina and I had submitted our paper to *Nature* announcing the first known embryos in arthrodiran fishes and demonstrating that the basic structure of the arthrodiran pelvic fin bore remarkable similarities to those fishes that mated using claspers. We suggested that copulation was far more widespread in early fish evolution than previously thought. By late January 2009 we were again in print, this time along with a large news and views section written by Professor Per Ahlberg of Uppsala University on the significance of our finds, a profile piece on the first author (me), and an online mini-documentary including interviews with all the authors. In essence, they had given us the works. The press reaction to the story was huge, and the media frenzy played out all over again for the members of our research team.

We had found the smoke, but not the gun: the male arthrodire clasper, the world's oldest vertebrate willy, continued to elude us – or so we thought. It took an interested outsider to recognize what was right in front of us.

6

Finding the Daddy Fish

*The greater part of the year 1920 was spent in the
electrical stimulation of pithed skate ...
I have not yet met with skates in copula, but a
reliable authority informs me that in these larger
species only one clasper is inserted at a time.*

William Harold Leigh-Sharpe (1881–1950)
'Memoirs on Claspers' (1920–1926)

Sometimes, when immersed in research, we are fortunate
to serendipitously uncover rare glimpses of the past lives
of extraordinary people, led there by the need to find out
information on very obscure topics. Whilst investigating
the internal anatomy of the male genital organs of sharks
and rays so we would know what to look for in the search
for our missing arthrodire clasper, I stumbled across a

series of scientific papers that provided me with all the information I could possibly want about their private parts. Published in the *Journal of Morphology* between 1920 and 1926, the papers were by a truly remarkable and little-known English scientist named William Harold Leigh-Sharpe. The most extraordinary thing about his work – a beautifully illustrated set of anatomical dissections combined with the results of his experiments showing how the claspers of the chondrichthyan fishes (cartilaginous fishes which include sharks and rays) functioned – is that it was completed by a scientist whose lifelong speciality was actually not sharks, nor even vertebrates, but the study of parasitic copepods – minute little crustaceans distantly related to crabs and lobsters. Leigh-Sharpe became fascinated by how the parasites were getting into the sharks' bodies, so he went off on a tangent for several years to investigate the anatomy of the claspers. He became the first zoologist to determine that sharks used sea water to pump their sperm through a siphon gland (his experiments involved flushing water from a hose through a dead shark's claspers) and thus introduced parasites. His research ultimately led him to determine that Cowper's glands in humans (which produce a component of semen) actually perform the same function and were developed in the same way as the clasper glands in sharks. In sharks this gland secretes a fluid rich in proteins which coagulate immediately on meeting sea water to close up the groove along the clasper and form a tube, thus allowing sperm to be effectively passed from the clasper to the female. The secretions also help lubricate the clasper and aid the passage of sperm through the clasper. Thus, as Leigh-Sharpe observed, it functions in the same way as does Cowper's gland in humans. Leigh-Sharpe left a lasting legacy of some 71 scientific papers on claspers, copepods and other animals, as well as his contribution to a major zoology textbook which was first authored by well-known British novelist and zoologist H.G. Wells. Outside the sphere of science, Leigh-Sharpe was also a gifted composer, publishing ten original pieces for the piano. He died quite penniless in 1950.

But how or why does one become utterly obsessed by shark claspers? For myself, as someone similarly fixated, it was about finding their prehistoric origins. Following our team's discovery of the long pelvic bone that supports claspers – the basipterygium – in ancient placoderm fishes, we next needed to find evidence that the claspers themselves really did exist in this diverse arthrodire group of extinct armored fishes. Our discovery of the last missing piece in the puzzle came swiftly after an opportune visit by one of the world's most eminent paleontologists, Professor Per Erik Ahlberg from Uppsala, Sweden. Per was an old friend of mine — he and I first corresponded back in 1989 when he was a postgraduate student studying the weird beady-eyed group of sarcopterygian (lobe-finned) fishes called porolepiforms. Per had also written a short commentary piece in *Nature* about my discoveries at Gogo in 1989, highlighting the 'incalculable scientific value' of the fossils due to their exquisite preservation. We finally met in person in 1992 on one of my visits to work at the Natural History Museum in London where he had just kickstarted his career as the new curator of fossil fishes.

Per visited me in Australia on two occasions over the 1990s, including on one memorable trip when he joined Dr Kate Trinajstic and I on a field trip to the remote dusty interior of Western Australia. A doctoral student at the time, Kate was studying the Gneudna Formation rocks and fossils near Williambury Station (150 miles or 250 kilometers inland from Carnarvon). Our mission was to collect vertebrate fossils and assist Kate in her task of mapping the general geology of the formation. Two Russian paleontologists who were accompanying a traveling dinosaur exhibition also came along, and an action-packed trip it turned out to be.

It started with Per, Kate and I arriving at the station shearers' quarters to find that it was 'national geologists day in Russia', which the Russians insisted we celebrate by sharing numerous toasts of neat vodka with them. Sore-headed but intrepid, the next day we all ventured out into the field and began collecting fossils in earnest. After a couple of days of toil the skies opened up and it poured heavily; in fact we just managed to escape from the field site in our four-wheel drive before the track turned into a sodden and muddy mess.

With the heavy rain, the snakes and other nasty venomous critters headed for high, dry ground, in this case the station house and shearers' quarters where we were staying. The station owner, Josh Percy, was bitten by a particularly nasty kind of large yellow and red centipede. Also a trained nurse, Kate offered her help and began monitoring his vital signs. Normally such outback emergencies would result in a call to the Flying Doctor Service, but the teeming rain had made the landing strip unusable. It was touch and go for few days, but in the end the swelling in his leg subsided and he began to show signs of recovery.

Per came back to visit Perth again in 1997 for a vertebrate conference and also accompanied us, along with a host of other international paleontologists, on a field trip to the Gogo fossil sites. In recent years he has became involved in projects involving the Gogo fish fossils, which mission again led him halfway across the world in early 2009 to examine Gogo specimens in the Western Australian Museum, a good number of which hadn't even been prepared.

One of the specimens Per and Kate, who was based at Curtin University in Perth, examined together was an arthrodire, WAM 03.3.28, which showed excellent preservation of the pelvic girdle. I had found it in the field back in 2001 and then assigned it as part of a project to a zoology honors student, Kate Bifield, primarily to examine the structure of the tail and pelvic fins. For many years only cursory study had been given to the tails of the Gogo placoderms, mainly because few were ever preserved to any large extent. The most detailed work previously published on the Gogo arthrodires had been by English zoologist Dr Kim Dennis-Bryan, whose speciality was in mammals, but who assisted Dr Roger Miles with some Gogo arthrodire specimens back in the late 1970s. Pretty soon Kim took over and became the main expert on these fishes.

Kim Dennis-Bryan's work describing the Gogo placoderms with Roger Miles provided the most detailed studies of the tail and pelvic girdles in any of the Gogo specimens. Their account of this area in *Incisoscutum* is particularly precise, giving clear anatomical descriptions of the vertebral elements, the pelvic girdle and the radials, which are the bony elements supporting the fin. At the back end of every Gogo arthrodire pelvic girdle was a large hole that they labeled the metapterygium, or rear part of the pelvic fin. Our recent discoveries had clarified that this was in fact an articulation facet for the basipterygium – an important distinction, remembering that the basipterygium is a long, rearward-facing bone in the pelvic fin of all sharks, rays and holocephalans that joins to the claspers in males.

So back to that day in early 2009, when Per was visiting the Western Australian Museum and re-examining our arthrodire specimen, WAM 03.3.28, with Kate. Per spotted something that had for years eluded our own eyes. What we had thought was the inner side of the opposite pelvic bone to the large one exposed was in reality the missing piece of our puzzle. Lo and behold, it was the basipterygium which had the actual clasper fused onto the end of it: he had found what he dubbed a fossil 'todger'. The clasper was highly unusual, and this had disguised its presence. What we were looking for was something that resembled the claspers found in ptyctodontids or modern sharks, a separate bony or cartilage element that articulated with the rear end of the basipterygium. Instead, this clasper was fused to the basipterygium, making it appear like a conical head with well-demarcated ridges and irregularities on its knobbly head – the business end, if you like.

Though not strictly homologous (or corresponding) in the scientific sense, the knobbly end of the arthodire clasper reminds one somewhat of the head of the mammalian penis. The ridges and irregular bony structures would have helped hold it in place during the act of copulation, signaling that maybe it wasn't as erectile as modern shark claspers. The fact that it was fixed upon a long, sinuously curved hollow shaft of bone means there was no part of the structure that could easily move relative to the fixed position of the pelvic fin. There was also no obvious distinct groove for transferring sperm along the structure, so this function must still have been carried through the soft tissue attached to the basipterygium via a tube or groove.

Finding this male copulatory organ of the largest

group of placoderms, the arthrodires, meant that if our hypothesis was correct, there should also be a female counterpart to demonstrate that these fishes were indeed sexually dimorphic.

It did not take us long to find another example of a basipterygium in the marvelous suite of fossils that Kate Bifield had prepared to study pelvic girdles. The second example was smaller and lacking the extreme length seen on the male basipterygium. Unfortunately the tip of the specimen was not exposed, being tucked under the pelvic girdle, so we were not able to describe it further or confirm that it lacked the knobbly end as seen in the male organ, but it was shorter and broader than the long male clasper, which made us think this could only be the female expression of the basipterygium.

If we consider how modern sharks and their kin copulate it helps us understand the physiology of placoderms, and how their mating might have differed. Claspers in modern sharks are extremely complex organs, and even more so in rays and holocephalans. For a start, they are true erectile organs, which means blood must be pumped into the spongy tissue to inflate the clasper. When the mood to mate takes (some of) us mammals, we approach the female with penis already erect, visibly showing our intention or desire to copulate. In sharks and rays the clasper is first partly erected to move it forwards, then it is inserted into the female's cloaca and further inflated. Once inside it cannot be easily dislodged while the vital act of transferring sperm takes place.

To facilitate this task, as mentioned in Chapter 3, some sharks have many small hooked and spiny scales on the ends of their claspers, so that when erect they grip the inside of the female's body cavity. In female sharks the ovaries open directly into the body cavity, or coelom; there's no vagina or special reproductive tube to receive the claspers. It means there is a lot of open and rather loose space that the clasper penetrates once it has entered the female's cloaca – anything to anchor it in the female's body while he transfers the sperm would obviously be a big evolutionary advantage.

In modern-day grey reef sharks (*Carcharhinus amblyrhinchus*), pre-mating rituals take up a lot of time as males first vie against each other just to get close to the females, then bite them about the head, neck and back until they succumb and the males can hold onto the pectoral fin with their mouth. The male turns the female over and finds a place where he can pin her down so that he can get purchase and insert the clasper. The clasper rapidly rises forwards, is quickly inserted into the female and remains there at least four to five minutes whilst sperm is transferred to the female in little clumped tangles called 'spermatozeugmata'.

Dr Brad Norman of Murdoch University in Western Australia has spent his life studying the gentle giants of the shark world, the tropical filter-feeding whale sharks (*Rhinodon typicus*). He and a team of colleagues led by Dr Jennifer Schmidt recently discovered that female whale sharks can store sperm and use it after mating, fertilizing successive embryos over the course of a year. Even more amazing is what mega-mammas these fishes can be: one pregnant female was recorded with some 300 embryos inside her.

Once the eggs are fertilized, the females of all sharks and rays either develop embryos internally and give birth to a litter of pups (viviparity), or internally encase them in hard keratinous egg cases, lay them, and leave them to fend for themselves (oviparity). In some species the fetal sharks growing inside the mother mature to the point where they start to eat each other for nourishment, leaving only a couple of large pups to be born. In rare cases some sharks have reproduced by means of parthogenesis, whereby the female develops fertile eggs that are her genetically identical clones, without fertilization by a male. This has been observed in several species of sharks in captivity where the female was not in contact with a male in her whole life, but was still able to bear live young. In one recently reported case even these offspring were able to reproduce successfully, showing this was not some kind of dead-end evolutionary birth mutation.

Our research had journeyed from finding embryos in two major groups of placoderms to identifying a time in early vertebrate evolution when internal fertilization first appeared, and then rapidly became common. Why should this have occurred? Robert MacArthur and Edward O. Wilson's 1967 ecology classic *The Theory of Island Biogeography* determined that selection pressures drive evolution into two general directions: r or K strategies. R strategies occur in unstable or pressured environments where the need is to reproduce quickly. Animals respond by producing vast numbers of offspring, as the cut-throat world they live in means most of them will become prey for other organisms; so long as a few survive to reproduce, their species will continue. Examples of r-strategy fishes are cod, which produce millions of viable young but only a few will reach maturity.

In stable and predictable environments, the K-strategy organisms invest more time and energy to produce fewer but more developed offspring. These species include most mammals, we humans in particular. Our discoveries of placoderms nurturing embryos thus identified the first K-strategists in the vertebrate line.

Does this imply they lived in one of the earliest kinds of stable environment? It might very well indicate that the reefs on which these fishes lived in the warm, tropical seas of 375 million years ago had a degree of ecological balance to them. For fishes to invest in rearing just a few young to large size and not be eaten themselves while doing so suggests that there were predictable micro-environments that they could find respite in. Perhaps the evolution of large reef systems provided the many nooks and crannies for pregnant fish to be safe inside, away from much larger predators. As yet we have only noted pregnancies in small species and know very little about the reproductive biology of the largest placoderms, such as the 25-foot (7.6-metre) long predatory *Dunkleosteus*, except that an egg case attributed to this species was reported at a 2010 Society of Vertebrate Paleontology meeting by Dr Bob Carr of the University of Ohio. The large egg case with small tuberculated bones inside suggests that the very biggest placoderms must have had internal fertilization, otherwise they would have laid masses of very small eggs as other fishes do. Such large shark-like egg cases reinforce the idea that these ancient arthrodires also had a shark-like reproductive system which used both viviparity and in this case oviparity, whereby the young are developed internally to an advanced stage then laid as a few large eggs.

This latest discovery courtesy of Per, coupled with more observations and speculations on shark reproductive behavior, prompted a number of questions concerning how the first jawed fishes mated and might have nurtured their young. Looking at the scientific literature, it seemed very few people had ever seriously pondered the question that fascinated us now: how did the ancient armored placoderms have sex? Fewer still would have entertained thoughts of what positions they may have favored.

But the speculative realm of placoderm pornography is exactly where the next stage of our research took us, specifically to make a computer animation presenting our best guess as to what the earliest known act of vertebrate love may have looked like. (Trust me, I'm a scientist!)

7

Down and Dirty in the Devonian

Lechmere had been wrecked on North Island,
New Zealand, and had been living very happily
amongst the Maoris who had rescued him, roaming
the hills and forests, collecting specimens, and
thoroughly enjoying himself. Although this was the
time of the Maori wars, they treated him with great
hospitality, and to the end of his life he loved to
talk of his adventures with them, and to display the
tattoos on his back — of various designs, including
a sailing canoe, and on his wedding finger — a ring!
He had left only just in time, he declared, to avoid
marrying the chief's daughter!

Yseult Bridges
Child of the Tropics

In 1856 when amateur British naturalist Robert John Lechmere Guppy was stranded on the coast of New Zealand, living with the Maoris, he would never have guessed that his family name would live on not through his heroic deeds and adventurous travels, nor by the numerous scientific papers he would write in later years, but mainly due to one single deed of some years previous. That deed was to send to the keeper of fishes at the Natural History Museum in London specimens of a small colorful fish from Trinidad which would eventually bear his name, *Giradinus guppii*, from which the common name 'guppy' was derived. Much later, after the popular term guppy for the fish had been firmly established, the name was found to be 'superseded' by the name *Poecilia reticulata* – meaning it had actually been assigned earlier than *Giradinus guppy* and thus had naming precedence.

Today guppies are a very popular fish kept in aquaria around the world. Unlike most of their ray-finned kin which spawn in water, guppies, along with a few other groups of ray-finned fishes, actually mate through sometimes strange but intimate acts of fertilization. Yet none of them have anything like the well-developed male claspers seen in sharks and the ancient armored placoderms. So in order to make a best scientific guess at how placoderms might have had sex, we first needed to take a look at the broader spectrum of fish fertilization, to see just how funky fishes can get when they do the deed.

Fish have a variety of clever adaptations that have evolved independently in different lineages to overcome the inherent dangers of spawning in open water. (Spawning in a calm ocean is fine, but in swift-flowing streams and rivers, any novel way of keeping sperm and egg closer together would be advantageous to survival.) Hence we find a number of special mating adaptations have emerged in fishes that, over the course of their evolution, have moved from the marine realm into freshwater habitats.

Guppies are a prime example. The males are equipped with a modified anal fin spine called a 'gonopodium' which has a groove in it that can pass sperm into the female's cloaca. Tropical fish hobbyists around the world delight in breeding guppies, though it's not exactly rocket science: just putting them in a tank with the right temperature and right water quality will do it.

Other tropical ray-finned fishes have truly bizarre sex lives, foregoing copulation for an act of rapid oral sex to fertilize their eggs. A paper published by Japanese researcher Masanori Kohda and his colleagues has the intriguing title 'Sperm drinking by female catfishes: A novel mode of insemination'. A novel way to get pregnant indeed, even if somewhat hard to swallow. (Female readers needn't fear falling pregnant the same way – it only works if you are a small Amazonian catfish.) The bronze catfish, *Corydoras anaeus*, a popular r-strategy aquarium species around the world, has evolved a mating ritual whereby the females align their bodies at right angles to the middle of the male, with their mouths close to the male's genital opening. In a split second the male ejects his sperm which is sucked inside the female's mouth and almost immediately she sheds her now fertilized eggs out of her cloaca.

At first scientists couldn't figure out how the system worked so quickly, but by placing blue dye in the water they were able to watch the flow of male sperm along with swallowed water. Once in the female's mouth, the sperm was transferred by various ducts into the body cavity of the female, where the eggs had been previously deposited from the ovaries. A perfect system, bearing in mind that these tiny fishes live in fast-flowing streams where sperm shed in water intended for eggs might be quickly washed away with the currents.

When the male of the deep-sea angler fish *Ceratius* meets an attractive female angler, it's usually not with flowers and chocolates tucked under one fin. It is deep in the cold, dark, abyssal depths of the ocean, as far down as 3000 feet (900 meters). There in the pitch-black waters, it's not easy to spot a mate, let alone a pretty one. And like all males seeking a mate, the male angler fish has to make a very big decision when he finally makes contact with a prospect. Will he give up the free swimming roustabout life of an independent fellow, able to swim around and feed with his mates whenever he feels like it? Or will he attach his tiny body to the female – who is maybe 50 times his body mass – to eventually fuse completely with her, slowly relinquishing all his internal organs until almost nothing is left of him but an outer skin and his big bag of testicles, and thus serve his only purpose: inseminating her eggs?

In essence the female angler fish becomes a kind of hermaphrodite, and the male a parasite, always on hand to fertilize her eggs. Despite this less than appealing existence where he will lose all aspects of his identity apart from his essential ability to create sperm and shed it into the female, something inside, preset by his genes, urges him to attach himself to the available female, as

Down and Dirty in the Devonian

he might never meet another one in the dark, gloomy waters of the abyss. Indeed, such parasitic males don't even enjoy the luxury of monogamy, as the much larger female fish will occasionally take on additional males to provide a range of genetic variability each time her eggs are fertilized.

In some cases where fishes don't actually copulate, the sexual behavior can still be quite radical. When the females of the bluehead wrasse (*Thalassoma bifasciatum*) head out with large males to the spawning site, it pays for them to be wary of the masses of puny males that relentlessly harass them from the sidelines. These runts regularly go up and touch a female, on occasion causing her to shed some of her eggs. The frenzied group of runt males then shamelessly shed their sperm to mass-fertilize the rogue eggs.

Other forms of multiple sex matings in fishes are seen in capelins (*Mallotus*), which live around the Arctic waters of northern Europe, Russia, Iceland and Canada. These small fishes form large shoals to become the dominant filter-feeding animal in Arctic waters and an important food source, so their mode of reproduction is of great interest in northern countries. Some mate in the tidal zone in less than a few feet of water, while others make it all the way up the beach. Sometimes intimate threesomes are formed, where two males and one female will get together for a high-tidal zone quickie, the males squeezing the eggs out of the female during each frantic run onto the beach with each washing wave, kind of like a female caught in a male sandwich. Once their sperm has been shed onto the eggs, the vast majority of the males then die. (Further examples of this kind of behavior are described in Chapter 10 for grunions, a closely related species.)

So when it comes to mating, fish can get funky in many bizarre ways. Most people would expect fish to breed opportunistically with whichever mate they can find but this is not necessarily so – there are nearly always exceptions to any rule in biology. Seahorses (*Hippocampus*) and some pipefish species (*Syngnathus*) are monogamous for the span of the breeding season, behavior thought to have evolved as a way of guarding the mate and thus having a better chance of breeding success.

The seahorse is also unusual in its mating behavior and method of parental care. After a lengthy period of courtship, which can involve days of dreamy side-by-side swimming, changing colors and wrapping around the same strand of seagrass together, they climax in a sensual dance in which the couple are intertwined for up to eight hours. During this time they eventually take up a position with their snouts meeting, spiraling upwards in a slow drift. The male seahorse inflates his brood pouch on the front of his body by blowing water into it, showing the female he is ready to receive and fertilize her eggs. When the female deigns to deposit her eggs, he instantly fertilizes them with a carefully timed emission of sperm from within his pouch. Having been effectively impregnated, the male then takes care of the fertilized eggs in his pouch, protecting them from danger and making sure he provides a suitable environment for the eggs to hatch in, which may take anywhere from nine to 45 days. Once the young seahorses emerge and disperse, his fatherly duty is completed.

Bearing in mind the strange range of mating behaviors that can be witnessed in living fishes, the next question is how would our ancient placoderm fishes have mated? The ptyctodontids may have used their curved, heavily spiked claspers to either insert into the female or simply to grasp the female and force cloaca-to-cloaca contact. But the fact is that we have no actual evidence that the clasper was inserted into the female, and on closer examination it looks downright difficult to do with a curved bony structure that ends in hooks and sharp ridges.

Some fossil ptyctodontids, such as the *Cyrtacanthus* found in the Late Devonian Cleveland Shale of Ohio, have huge claspers – around six inches (15 centimeters) long. These hook-shaped elements with sharp rows of spines simply don't look like they are made to insert inside the delicate insides of any female, nor is there any visible trace of a groove that could be used to transfer sperm. Perhaps ptyctodontids, like our *Materpiscis* mother fish, were indeed having primitive sex, but through a more basic cloacal-kissing, much the same way that many frogs and birds mate. By using the hooked clasper to grasp the female and hold their bodies close together, the male could direct his sperm into the female and thus fertilize her eggs internally.

But the *Incisoscutum* clasper that Per Ahlberg first identified on his 2009 visit to Western Australia represents a different situation entirely. Here we see a very long (hint as to its real function), relatively straight (not hook-shaped) and slender shaft on the basipterygium, with a small but slightly knobbly clasper component on the business end. The length is a hint to its real function. The orientation of the clasper is facing away from the head of the male fish, and the degree of flexibility is clearly quite limited as can be seen by the large, vertically oriented articulation facet – this clasper could have easily been inserted deep inside the female's cloacal opening. And as the shaft is smooth and slender, it most likely had soft tissue enveloping it, and thus could well have contained a groove for the transfer of sperm. Add to all this the fact that the structure appears not to be useful for grasping a female like the hooked and knobbly ptyctodontid clasper and we see here the first example of what could be a true intromittent (designed to be inserted) organ – perhaps the world's first such prototype in a backboned animal, and not too far off the actual design of the human penis in shape.

If this is the case, placoderms like *Incisoscutum* were indeed well-endowed, with a much longer clasper than the average mammalian penis (relative to overall body size), as necessitated by the awkward clumsiness of the thick bony armor plates that covered their bodies. The enormous penis of the nine-banded armadillo (*Dasypus novemcinctus*) might be a similar case, where heavy body armor gets in the way of easy romance, so a larger male member is needed to compensate. If armadillos were human-sized, their penises would be the equivalent of about 4 feet (1.2 meters) long.

But the most amazing feature of this first proto-penis in the vertebrate world is that the clasper of *Incisoscutum* could only face backwards, away from the head of the male fish. This means that there is no way a simple 'missionary' style position could be assumed for mating, as in many living sharks and rays. Quite the opposite, in fact; mating must have been in an inverted '69' position, with the female on her back on the soft seabed floor whilst the male pushed the slightly erected clasper backwards into her cloacal opening.

How do we know the clasper was slightly erected? Well, the joint between the basipterygium (bearing the clasper head) and the pelvic girdle would have been flexible to some degree, so the need to make the tissue around it more erect for controlled insertion of the clasper seems likely. The male may also have used his jaws and his flexible pectoral fins to grasp parts of the female's armor and facilitate the twisting and writhing necessary to insert the clasper inside her.

Having sufficiently imagined how these ancient fish could have started doing the deed, we set about making our paleoporn movie. With no life-sized model of

Incisoscutum handy, we cast a model of the Gogo fish *Latocamurus* in the lead stud role as a very close substitute. After scanning the model to produce a 3-D computer grid outline of the fish's shape, the image was enhanced by hand to create a male with a long trailing clasper and a female with simple pelvic fins. We also played around with the colors to make slight differences between males and females, but that was really only scientific guesswork for art's sake.

The final clip showing the actual mating of the fishes involved the female lying on her back on the sea-floor; we figured placoderms would have a hard enough time of it trying to insert the clasper in mid-water. As in typical porn films (or so I'm told), the male in our production doesn't go in for elaborate courtship rituals. This was not because we didn't think placoderms had courtship rituals, just that we knew nothing about them (behavior-wise) and we were on a rather tight budget – these short CG animations cost a packet out of our research funds and we had to spend wisely. The video shows the male descending lustily upon the readied female lying on her back with pelvic fins spread widely. The male backs up and inserts the clasper rearwards into her waiting cloaca, depositing his sperm and then hastily wriggling out and swiming off. The female is seen lingering after him, eyes glistening longingly as he skedaddles away, possibly wondering to herself, 'Will he ever call?'

Our video was a good, educated guess, but who really knows what the first act of sexual intimacy between our deep-distant ancestors was like? All we can do is make up scenarios based on the available scientific knowledge. Yet this leads us to another question: when and where did sex first evolve in animals, and why? In order to tackle this series of questions we will need to poke around with fossils from much further back in time, to not long after the very origins of life itself.

Down and Dirty in the Devonian

8

At the Dawn
of Archaic Sex

Birth, and copulation, and death.
That's all the facts when you come to brass tacks:
Birth, and copulation, and death
I've been born, and once is enough.

T.S. Eliot
'Sweeney Agonistes'

So when and why did organisms first start reproducing by sex? By 'sex' here we mean that instead of simply shedding or budding off a piece of themselves to create a new identical form of life (a clone having the exact same DNA as the original), two organisms decided to get together and share their genetic material to create a more genetically variable kind of offspring.

Most primitive forms of life can be divided into those that lack a nucleus in their cells (prokaryotes) and those with a nucleus (eukaryotes). Most organisms that we associate with today – the complex multicellular animals and plants (called 'metazoans') – are eukaryotes which contain nuclear DNA inside them. Most eukaryotic organisms undergo sexual reproduction, the sharing of genetic material to make a new generation, although some can reproduce asexually by budding off identical clones. Some animals like hydras (a relative of the jellyfish) can reproduce sexually or bud off clones depending on food availability. 'Sex' in the biological sense is really defined by the process of meiosis and gametogenesis, when cells divide in such a way to produce gametes (for example, egg or sperm cells) by halving their chromosomes. When male and female gametes from different organisms unite, the chromosome halves recombine to begin making a genetically unique new organism. So how can fossils shed light on such microscopic and delicate processes that took place not just million of years ago, but probably a billion or more years gone by?

To try to answer this question, it might help to briefly look at the life of a truly extraordinary man named Reginald Sprigg, whose work in the field of geology begat a whole new field of study that has since revolutionized our understanding of the early evolution of multicellular organisms. Born in 1919 in Stansbury, on South Australia's Yorke Peninsula, Reg collected fossils and shells from his local beach as a boy and later become fascinated by minerals through a chance meeting with an old miner. Studying science at Adelaide University, he was fortunate to learn under greats such as Sir Douglas Mawson and Professor Cecil Madigan, both veterans of Antarctic exploration. Mawson is quoted as saying that Sprigg was his 'best ever student'. Sprigg was inquisitive and liked to question and challenge the views of his professors at time when it was not common to do so.

After he graduated in zoology in 1941, with honors in geology, Sprigg was brought on board an Australian Government top-secret project searching for Australian uranium deposits. This was wartime, and a great race had begun to develop and utilize the secret properties of uranium. It would result in the first atomic bombs being built, and ultimately end the war in the Pacific through the horrific events at Hiroshima and Nagasaki. Reg worked on several key deposits in Australia and was sent to study uranium deposits in the USA, Europe and the UK in order to extend his knowledge of the geological settings of uranium-bearing ores. On his return to Australia in 1950 he was perhaps the world's most highly regarded source on the subject.

Despite his groundbreaking work, forces within the Australian Government hindered his work, keeping information from him. In the end he handed his uranium studies over to others and switched his brilliant mind to working on petroleum exploration. But a discovery he stumbled upon during his uranium field work in the Ediacara Hills of the South Australia's Flinders Ranges remains a legacy. At the time he determined that the strange fossils were probably of Early Cambrian age (about 540 million years old), both through his geological mapping of the area and also because at the time no large metazoan (multicellular) fossils such as the ones he had found had ever been found in the older Precambrian rocks. Reg identified the fossils as being impressions of jellyfish and first exhibited them at an ANZAAS (Australian and New Zealand Association for the Advancement of Science) meeting in 1946. Recognizing the significance of the very old age of these finds, he published two important papers in the *Transaction of the Royal Society of South Australia* in 1947 and 1949, describing various species of early jellyfishes from his new sites.

Enter Martin Glaessner, Bohemian-born paleontologist extraordinaire. Trained in Vienna, Professor Martin Glaessner had fled Nazi Germany with his Russian ballerina wife during the war years to find work in New Guinea with Shell before ultimately arriving in Australia. After a stint in Melbourne he settled into a steady job at the University of Adelaide, where the Ediacara Hills fossils found by Reg Sprigg caught his attention. In 1958 Glaessner published a paper similar to Sprigg's on new forms of Lower Cambrian (that is, lower in rock strata) fossils from Ediacara, but in that year everything would change when a discovery from the other side of the world would show the Ediacaran fossils in a whole new light.

When Dr Trevor Ford from Leicester University published the first account of undoubted Precambrian-age fossils from a site in the Charnwood Forest of England in 1958, he described a frond-like organism he named *Charnia masoni*, similar forms of which were known at Ediacara. Martin Glaessner then went to print with his landmark paper in *Nature* on Precambrian jellyfish and other coelenterates (the phylum which jellyfish and sea-anemones belong) from Ediacara, Africa (Namibia) and England, announcing to the world that the oldest known assemblage of fossils came from Australia. This was soon followed by an article about these ancient fossils which nabbed him the cover of *Scientific American* in 1961. The Ediacaran fossils have been studied and collected intensely ever since, and still they keep shedding their secrets. Today these fossils are known as the 'Ediacaran Biota' and are accurately dated at around 560 million years old, well before the explosion of life occurred in the Cambrian period 540 million years ago.

Reg Sprigg's legacy lives on in the naming of a new geological time period, the first one to be described in over a century. The Ediacaran period was formally established in 2004, delineating an age range of 542 to 635 million years ago. This is now widely known as the time when multicellular life first emerged in a variety of shapes and sizes. And this means that for such diversity to occur at this time in life, sex must have evolved.

The man who would discover sex in the Ediacaran fossil record first started collecting in the Ediacara Hills back in 1971, but would not realize his discovery until 2008. Dr Jim Gehling, a colleague and friend who is now curator at the South Australian Museum in Adelaide, began working on the Ediacaran fossils in 1971, tracing out layers containing the fossils in other regions of the Flinders Ranges. In 1972, he and fellow worker Colin

Ford found a remarkable new site where fossils showed what appeared to be frond-like organisms with the broad base of the animal that held it to the sea-floor in the same bedding planes. These fossils challenged previous interpretations by Glaessner that the Ediacaran fossils were washed-up remains on intertidal flats. Gehling's work hinted that they could be much deeper water dwellers. The debate about Ediacaran organisms, what they actually are and how deep they lived goes on to this day, and one very recent discovery made international headlines when published in *Science* in 2008: it announced the discovery of the origin of sex.

The paper by Mary Droser and Jim Gehling described a new kind of organism from the Ediacara site which they named *Funisia*. *Funisia* was a worm-like tubular organism, the fossils of which are found abundantly at the Ediacara sites, so much so that different stages of its growth can be studied and measured in detail. Droser and Gehling identified that these organisms were budding off 'sprats', or juveniles from the adult, which were all at a similar growth stage. Hence instead of shedding or budding asexually (shedding identical clones of itself), as expected for primitive organisms of this period, it is likely that the 'sprats' developed all at the same time due to an act that begat them all: sex. Put simply, if they were budding asexually then a wider range of sizes would be expected in the juveniles. The fact that they were always at the same size suggested an act that was timed, a mutual shedding of sperm and eggs into the water as occurs for corals.

A *London Times* story about the discovery explained that 'the knobbly animal, named *Funisia dorothea*, is thought most likely to have reproduced in a similar way to modern corals and sponges, but little else is understood of its biology.' And, of course, the journalist went on to ask the scientists if the *Funisia* would have enjoyed sex:

'Sex for the creature would have been functional rather than a social affair,' Professor Droser, of the University of California, Riverside, said. 'I think they would have been way too basic to have enjoyed the sex. I don't think they would wind around each other. But I could be wrong – I would like to think they enjoyed it.'

These Ediacaran fossils provide circumstantial evidence, given the rigorous analysis of data carried out by the scientists, of a very early sexual reproductive event occurring in a similar way to how corals and sponges shed their gametes into the water before a period of new growth. This begs the question: could this form of sexual reproduction have been going on even further back in time?

The oldest known eukaryotic fossils are possibly the weird spirals resembling swirling party streamers known as *Grypania*, which have been found in rocks as old as 1.8 billion years in sites in both Michigan and Montana in the USA. One theory is that they are giant algae but others hold they could be large cyanobacteria, which in colonies build mounds of layered structures called stromatolites by trapping floating particles of sediment (excellent examples of these can be found alive today at Hamlin Bay in Western Australia). Bacteria do not use sex to reproduce; they just clone themselves. But it is more likely, given the rarity of large coil-shaped bacteria today, that the fossils are indeed algae, which all reproduce by sexual means, so *Grypania* could represent the oldest fossil evidence we have of sexually reproducing organisms?

Yet fossils can be more than just the remains of once-living creatures. Sometimes chemicals leave us traces of where life was before, like a ghost in the rocks. For example, in 1999 Jochen Brocks of the then Australian Geological Survey Organisation in Canberra and his colleagues pushed back the tentative origin of eukaryotes to about 2.7 billion years ago from their identification of complex biomarkers in the form of certain lipids (fats) in rocks of the Western Australian Pilbara region, which are unique in their chemical signature to those of living eukaryotic tissues. In August 2008 Birger Rasmussen of Curtin University in Western Australia and his colleagues published an important paper in *Nature* that critically reassessed the ages of biomarkers for eukaryotic cells. Their work shot down this earlier age of 2.7 billion years by chemical arguments that the biomarkers entered the rocks after metamorphic events – rocks being heated and crushed at great temperature and pressure. Their new estimates for the origins of reliable eukaryotic fossils now rest at 1.78 to 1.68 billion years ago, and this date, dear readers, is where we must currently park the idea of when sex first possibly began.

The age-old question that follows this is *why* did sexual reproduction begin? Why didn't life just keep evolving with simple cloning and asexual budding systems? Wouldn't it be easier if we were all like little freshwater hydras, where instead of performing complex mating rituals we humans simply grew a rather large lump on our bodies which eventually budded off like a festered sore, and from it emerged a perfect clone of ourselves? Maybe easier, but no fun at all, especially as we would all look the same

The Argentine lake duck, *Oxyura vittata*, sports the longest penis relative to body size of any vertebrate animal, this one measuring 1.3 feet (42.5 centimeters) extended. (Courtesy Dr Kevin McCracken, Alaska)

The author at the Gogo fossil site, in northern Western Australia in 2008. The rounded limestone concretions on the ground sometimes contain fossil fishes. (Peter Long)

An adult male Californian gray whale, *Eschrichtius robustus*, displaying his penis in San Ignacio Lagoon, Mexico. (© Michael S. Nolan/SeaPics.com)

jaws of embryo

The Mother Fish, *Materpiscis*, a 380-million-year-old fossil containing an embryo attached by a mineralized umbilical cord – evidence for internal fertilization. (John Long)

umbilical cord

clasper

The pelvic girdle and clasper of *Incisoscutum*, an arthrodire placoderm from Gogo. The long basipterygium has the clasper fused to the end of it. (Sexual organs arrowed.) (John Long)

basipterygium

A placoderm fossil from Victoria, *Austrophyllolepis*, showing long pelvic girdle (basipterygium), indicating it mated by copulation. (Sexual organs arrowed.) (John Long)

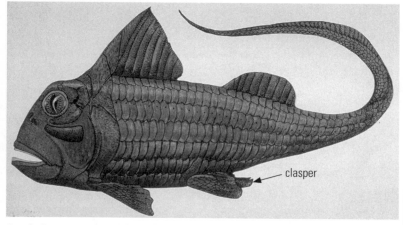

A male Gogo ptyctodontid fish, *Campbellodus,* showing clasper attached to pelvic fin. (Sexual organs arrowed.) (John Long)

A mating pair of nurse sharks, *Ginglymostoma cirratum,* in the Florida Keys, US. Sharks and rays all mate by copulation using claspers. (© Jeffrey C. Carrier/SeaPics.com)

You can see the mating scars on the flank of this female tiger shark, *Galeocerdo cuvieri.*

The grunion run, a mating frenzy of ray-finned fishes which mass-spawn under the full moon in summer.

A CT scan image of a remarkable 425-million-year-old fossil bivalved crustacean (ostracod) with penis preserved (arrowed), dubbed by tabloids as 'the world's oldest willy'. (Courtesy Dr David Siveter)

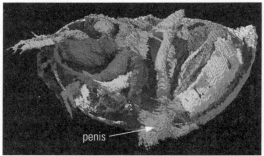

penis

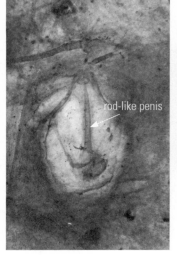

rod-like penis

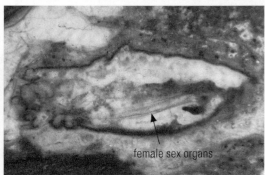

female sex organs

Male (left) and female (above) fossilized harvestman organs 410 million years old from the Rhynie Chert in Scotland. (Courtesy Dr Jason Dunlop)

(Top left) Two blue-back bedbugs having traumatic sex in which the male violently inserts his penis into the body of the female.

(Above) A female praying mantis mating with and eating her mate. The act of being eaten apparently causes the male to eject more sperm during his last moments.

(Left) Dragonflies were one of the first animals studied to show that males can remove the sperm of previous matings while inseminating a female.

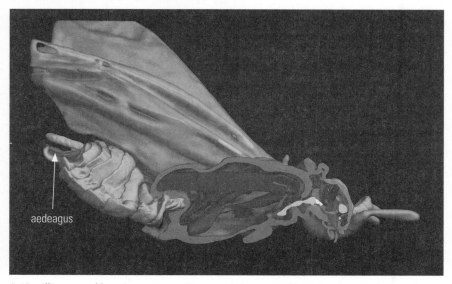

aedeagus

A 42-million-year-old stresipteran insect, *Mengea tertiaria*, preserved in amber, showing the insect's penis (aedeagus). (Courtsey Professor Hans Pohl, Germany)

Necrophiliac snakes. Four male garter snakes in Manitoba, Canada, courting (chin-rubbing) a female who is so long dead that she's turned blue. (Courtesy Professor Rick Shine)

Captive Galápagos turtles in the act of mating. It takes a few hours and a long penis for the male to reach the female's cloaca with large shells in the way.

A 70-million-year-old plesiosaur, *Polycotylus,* from Kansas, with embryo, the first evidence these animals reproduced by copulation. (© Natural History Museum of Los Angeles County, with permission)

Fossil dire wolf skeleton, *Canis dirus*, from La Brea site in Los Angeles. This example is a male clearly showing the penis bone (baculum). (John Long, with permission Page Museum, LA)

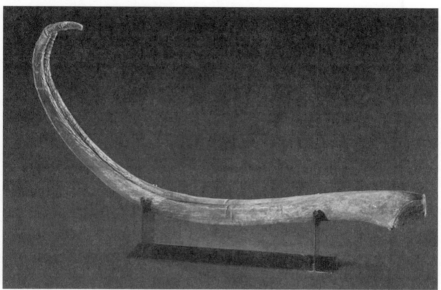

This fossil penis bone (baculum) of a 12,000-year-old walrus from Siberia measures 4.5 feet (140 centimeters) – the world's largest fossil sexual organ. (©2011 Ripley Entertainment Inc)

Spotted hyena mother, *Crocuta crocuta*, with her young pups. Female hyenas bear a penis-like clitoris and give birth painfully through it.

Bonobos, *Pan paniscus*, are our closest animal relatives, and like us they show a diverse range of sexual behaviors, used for social as well as reproductive needs. (Courtesy Vanessa Woods/ Bonobo Handshake)

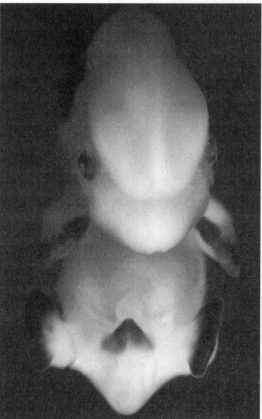

Image of a mouse embryo with Hoxd13 gene expression illuminated, showing the link between the genital area and limbs. (Courtesy Professor Martin Cohn, Florida)

and have the same personality traits. Imagine a world of just one person, multiplied a billion or more times. True it would make life easier for shoe and clothing manufacturers, but the first new disease through mutation to come along could potentially wipe the whole population out.

Aside from the social benefits, sexual populations have two main evolutionary advantages over asexual ones. Firstly, they can adapt more readily to changes in environment, and secondly, they are less prone to accumulation of deleterious mutations in their genes. British scientists Peter Keightley and Adam Eyre-Walker undertook some experiments that estimated genetic mutation rates in a range of animal species, but in particular focusing on fruit flies (*Drosophila*). They concluded that sex is maintained not just to purge the genome (the complete genetic material) of seriously harmful mutations, it is also principally driven by adaptive evolution, perhaps in combination with other mechanisms. In simple terms, sexual reproduction, and sharing different DNA, gives us a better ability to cope with the unexpected challenges in our environment that would otherwise wipe us out.

Sarah Otto from the University of British Columbia has written extensively about the evolutionary implications of sexual reproduction and rightly points out that, while enabling diversity, it is a costly exercise to reproduce sexually. The animal or plant has to find or stumble upon a suitable partner, risk sharing diseases, and becomes an easy target for predators during mating, sometimes even a target for the mate itself, as with praying mantises and some other invertebrate species. Sex is not an efficient way of sharing genes. When we mate sexually we share only 50 per cent of our genetic material with our partner, whereas asexually budding organisms have 100 per cent of their genetic material carried into the next generation. And Otto highlights what biologists call the 'cost' of sex, in that sexually reproducing organisms need to produce twice as many offspring as asexual organisms or they lose out in the population race.

Despite these drawbacks, evolution has shaped the living world in such a way that few large creatures today actually reproduce asexually (only about 0.1 per cent of all living organisms, excluding bacteria of course). Sex generates variation, and that is certainly a good thing when dealing with the constant and unpredictable changes in our environment: continents are slowly moving to new latitudes, ocean currents change, the climate shifts, or sudden traumatic events occur with volcanic eruptions or sudden (in geological terms) sea-level changes. Populations with genetic variability can adapt more readily to such pressures than those without much variation. The great German biologist August Weissman first said this back in 1889 and, despite much new work analyzing the pros and cons of sexuality, it still holds true today.

Once single-celled organisms began building more complex bodies (metazoans), sexual reproduction became the dominant method of reproduction. The explosion of life at the start of the Cambrian period, 540 million years ago, heralded the coming of many different kinds of animal body plans, most of which are still with us today. These include the first worms, mollusks (eg clams, snails) and joint-legged animals (arthropods, like insects, crabs and spiders). And this last group, as it turns out, enjoys some of the weirdest and most perverse sexual behavior of any creatures.

9

Sex and the Single Ostracod

*A violent but evolutionarily effective mating strategy
has been spotted in spiders from Israel. Males of the
aptly-named* Harpactea sadistica *species pierce
the abdomen of females, fertilizing their eggs directly in
the ovaries. The so-called traumatic insemination gives
the first male to inseminate a reproductive advantage by
bypassing structures in the females' genitalia.*

BBC News online, 30 April 2009

There is nothing more bizarre, brutal or twisted than
the sex lives of insects, spiders and other joint-legged
creatures, collectively known as the arthropods. From
bedbugs that brutally rape the females by stabbing their

bodies with a saber-like penis, and others that carry out homosexual rape to implant their sperm inside the competitor male's testes, the world of arthropod sex is truly horrific if compared with even the worst of our own mammalian perversions.

The arthropods include insects, spiders, mites, centipedes, crabs and other crustaceans, and a host of lesser known and extinct groups. Yet these creatures have evolved systems of reproduction that work well and have made them lords of the terrestrial environment of our planet. There are more known species of insect than any other group of land-dwelling life (some 1.178 million species are described, which equates to 80 per cent of all known animals). They also have an excellent and fascinating fossil record spanning the past 540 million years, possibly even extending a little further back if enigmatic Ediacaran fossils like *Spriggina* and *Parvanocorina* are one day shown to be arthropods.

Most arthropod fossils are found as impressions of their hard outer shell or cuticle, which in life is formed of a hard chitinous material. Many of us might recognize a trilobite if shown a picture of one – a segmented kind of fossil found commonly in rocks of the Palaeozoic era (540 to 250 million years ago). Trilobites represent the first great radiation of an arthropod group, with many thousands of different species appearing shortly after the group first evolved. Crustaceans are perhaps the best known of all edible arthropods, and they include crabs, crayfish and shrimp, as well as a broad range of other groups such as barnacles, copepods, isopods and amphipods. In fact, recent scientific study published in *Nature* of the DNA of a wide range of arthropods by Dr Jerome Regier from Duke University in North Carolina and his colleagues has proven that insects are now classified as a subgroup within the Crustacea group.

One crustacean, the humble barnacle, was the surprising subject of some of Charles Darwin's finest descriptive zoological work. Barnacles might look like immobile shell-covered creatures, but inside their hard exoskeleton is a little animal with jointed legs just like the rest of the crustacean clan. Yet barnacles display the most deviation of any group from the standard arthropod body plan, with many showing extreme forms of sexual dimorphism through widely differing male and female morphologies. One of Darwin's greatest discoveries was that some Cirripede or goose barnacles have tiny males that were parasitic on the females. This was a wondrous revelation for the time because other zoologists had thought the strange organisms inside the shells of the female barnacles were foreign parasites belonging to some other kind of animal group. Darwin accomplished an amazing set of descriptions of his barnacles, even measuring their tiny penises down to an accuracy of one thousandth of an inch (0.025 mm):

> In one case in which I dissected out the penis, I found it in its contracted state 41/1000th of an inch in length, equal to that of the entire capitulum and peduncle; in a specimen, in which the penis had been naturally exserted, the part which protruded (m [male]) was by itself rather longer than the whole animal; and as this specimen had been placed in spirits of wine, the organ no doubt was contracted; hence I think it probable that the prosbosciformed penis, when fully stretched out, would equal twice the length of the entire animal.

This description highlights the impressive relative size of the barnacle male penis, which in some species of goose barnacle can be eight or more times the length of the entire male, making it relatively the largest male copulatory structure of any animal yet known. So, if our

8-inch (20-centimetre) Argentine duck has a penis up to 1.3 feet (42.5 centimetres) long, a barnacle the same size would have a penis around 5.3 feet (1.6 metres) long!

Barnacles start life as mobile free-swimming larvae that eventually attach themselves to a hard rocky base at which point they develop their hard plate shell casings and live there for the rest of their lives, filter-feeding particles of food out of the water around them with their whiskery little legs. They are thus in no position to wander far in search of a mate. In order to reproduce, standard run-of-the mill intertidal barnacles are hermaphroditic, carrying both male and female organs in each individual. For such immobile creatures to inseminate one another, they need a flexible and, in some cases, extraordinarily long kind of penis. Once mating is complete, the barnacle sheds the penis and grows a new one the following year, just as deer shed their antlers.

Hey, guys: wouldn't it be handy if you could simply change the shape of your penis to suit the immediate job at hand? A recent discovery by Christopher Neufeld and Richard Palmer of the University of Alberta has proven that some barnacles can do exactly that. The species they studied can alter the shape of its penis according to the energy of the environment it inhabits. As they explain:

We observed that penises of an intertidal barnacle (*Balanus glandula*) from wave-exposed shores were shorter than, stouter than, and more than twice as massive for their length as those from nearby protected bays. In addition, penis shape variation was tightly correlated with maximum velocity of breaking waves, and, on all shores, larger barnacles had disproportionately stouter penises.

Those living on rough tidal zones tend to have shorter, more robust penises while those living in quieter waters develop the more elongated, gracile kind of male member. To get this kind of quantitative data, scientists must perform highly delicate experiments and take very accurate measurements. Neufeld and Palmer describe their approach to measuring fully erect barnacle penises as no easy task – unsurprisingly, once removed from their habitat and taken back to the lab, it's not easy to get barnacles in a romantic mood. Instead they are simply held down and made erect using a siphon of water pumped into the male organ, then measured and photographed. Their paper describes the procedure eloquently as follows:

The penis was oriented perpendicular to the field of view, and pressure was applied to the syringe to slowly inflate the penis until (i) the glue failed, (ii) the soma tissue or cuticle ruptured, or (iii) the penis inflated fully. Full inflation was recorded when additional pressure on the syringe failed to extend the penis further and all annulations of the penis cuticle disappeared. At this point, the penis was photographed again. This process was repeated for approximately 20 individuals per site until we had achieved full penis extension for three individuals from each population.

Their work had its moments when everything goes wrong, such as when the Krazy glue broke and a half-inflated barnacle penis went out of control in the lab. Mind you, for the barnacle it must have felt like an perverse alien abduction, but nonetheless such studies are important in determining how much of a role population genetics plays in developing such characteristics. In this case the researchers concluded that genetic plasticity was the main factor in enabling the barnacles to switch their penis shapes when the conditions they were living in changed, an effective adaptive strategy to cope with variable turbulence or flow conditions in the barnacles' environment.

Crustaceans not only hold the record for the largest known penis (relative to body size) of any animal, living or fossil, but also that of the oldest known well-preserved fossil male organ. One of the primitive groups of Crustacea that we do not eat – and, indeed, that most people have no knowledge of – are the ostracods or 'seed shrimp', so named as the animal lives within a bivalved carapace that from the outside makes it look like a seed.

Ostracods might look boring on the outside but inside their little shells they are hot bundles of seething sexual energy. Males sport two penises and in some species their tiny little testes contain enormous sperm wound up inside, up to six times the length of the grown male ostracod. Ostracods have an exceptional fossil record due to their calcareous bivalve shells (which preserve very well), and this has no doubt contributed to their triumph over all other animals as the world's oldest identified fossil penis specimen, as discovered by a British team in 2003.

Dated at around 425 million years old, the CT-scan imagery of this exceptionally well preserved fossil revealed a wide spectrum of soft tissues that had been mineralized within the cavities of the ostracod shell. The specimen was revealed to be a well-hung male, which was aptly named *Colymbosathon ecplecticos*, meaning 'amazing swimmer with a large penis'. (While it might sound like Johnny Weismuller, in reality there is no real resemblance.)

But back to serious business: how do well-equipped ostracods find a mate? A recent study of Caribbean ostracods showed that some of the males have quite elaborate courting techniques, using dazzling displays that involve luminescent shows designed to attract the non-dazzling ('photically silent') females. We have no way of knowing if our well-hung ancestral fossil ostracod used similarly spectacular tactics to attract its mates or simply relied on being the alpha male in the crowd. But if we combine information from both these strategies and use our imagination, we might envisage the *Colymbosathon* dangling its fluorescently glowing penis in the cool night waters to directly attract females – and female or not, it's hard to imagine such a device *not* attracting attention.

The harvestman, or daddy-long-legs, belongs to a group of long-legged spidery arthropods that are not quite spiders, but closely related to them in the arachnid group. The harvestmen, along with primeval mites and spider-like creatures called trigonotarbids, were amongst the first invertebrate animals known from fossils to invade land. This event took place in the late Silurian period about 420 million years ago, not long after the first primitive land plants had left the sea and colonized the near-shore habitats.

The same year the fossilized ostracod penis was announced, Dr Jason Dunlop, currently a curator at the Natural History Museum in Berlin, with his colleagues published a short note about some well-preserved fossil harvestman sexual organs in the journal *Nature*. From the famous Rhynie site in Scotland and dated at around 400 million years old, the fossils are thin cuticles preserved frozen in chert, a kind of rock that hardened from a soft sedimentary gel around the soft tissues and trapped them inside to show very fine detail of the fossilized parts. The male penis described is much like that of any modern daddy-long-legs; in fact, such creatures (like all spiders) use the front pair of appendages like arms to transfer sperm, so they are not 'penises' in the true sense of the word. Female sex organs (or ovipositors – which hold the egg) of the species were also found in the chert.

Recent research has found that, unlike male spiders, some of these daddy-long-legs are indeed good daddies. One study found that territorial males of the harvestman

Acutisoma proximum temporarily care for clutches that are left unattended by females and their harems. Indeed, back in 1990 Dr Gisella Mora of the University of Florida observed that one species of tropical harvestman, (Zygopachylus albomargini), actively guards the nest of fertilized eggs after mating has finished and the female has wandered away to forage. The males even guard nests containing the eggs of other females they didn't mate with. Dr Mora believes this to be unique in the arachnid world as the only clear case of male parental care behavior.

While much has been written about the mating behavior of insects, little can be deduced from their fossil record apart from the fact that ancient insects had similarly shaped bodies and mating apparatuses as their living counterparts. Fossil insects are sometimes found beautifully preserved in amber, the hardened fossil sap of ancient trees. One species identified in 42-million-year-old amber, Mengea tertiaria, belonged to the enigmatic fly-like group called stresipterans. A CT-scan study of the entire animal in amber revealed the remarkable internal anatomy of the little insect, even showing fine details of its aedeagus, its penis-like male organ.

So while we can't read too much into the possible sexual habits of fossil insects, living insects provide us with some of the most bizarre mating behaviors ever encountered. Although most insects reproduce sexually, some are able to clone off exact replicas of the adult; this is known as 'parthogenesis'. When insects do copulate the male generally achieves this by using his aedeagus to transfer sperm. This penis equivalent is part of the overall phallus situated at the end of his abdomen, which might also contain 'valvia' that help hold the aedeagus to the female whilst mating. When the male mates with the female he deposits sperm as packages (spermatophores) directly into her ovipore (egg duct) via a common genital opening (the bursa copulatrix) which forces the sperm into a storage chamber called the spermatheca. There the sperm is nourished from the spermathecal gland and also prepared for the job at hand by dissolving the hard external coat of protein covering them.

Most insects that reproduce like this go about it in a fairly standard way. You may have observed dragonflies stuck together by their tails in the act of mating, and that pretty well describes the process (although what goes on inside the female dragonfly is another matter, and is discussed later in this book when the topic of sperm competition is explored). Other insects use more traumatic variations of sexual intercourse to maximize their mating success, and this is where things get ugly.

Praying mantises are well known for their sadistic mating habits, commonly termed 'sexual cannibalism' whereby basically the female will eat the male during the act of copulation. While not a great result for the individual male, this actually has added benefits for the species: the female gets a nourishing meal that helps to feed her newly fertilized eggs, and the male is bizarrely stimulated to have a stronger ejaculation through the trauma of being eaten and ensuring his offspring will be numerous. In some cases, where female praying mantises have been previously well fed, they can mate without the male being eaten. The Chinese mantis uses elaborate courtship dances that precede mating when the female has been well nourished, but she still often eats the male thus getting dinner and a show.

Perhaps the nastiest case of insect sex is in the world of bedbugs (*Cimex*) and their related kin; their annoying bloodsucking bites seem like nothing once you realize what they are doing to each other. The male bedbug mates with the female through 'traumatic insemination' – literally by stabbing her in the abdomen with his knife-edged aedeagus to ejaculate directly into her wounded body (if you think this sounds like a scene from Andy Warhol's *Flesh for Frankenstein*, you are not far wrong). Insects lack the separate blood-lymph circulatory systems that higher animals do, so the sperm enter the combined circulatory system and eventually end up in the ovaries to fertilize the eggs. The wounds often heal, but not always, so evolution has clearly not yet perfected this bizarre way of mating. Sometimes the males even traumatically inseminate other males, possibly to displace their sperm for future matings, although there is no tested observational evidence to support this idea as yet.

A recent discovery of a spider in Israel that commits a similar traumatic mating ritual indicates that this brutal kind of reproduction must have evolved independently a number of times. The male *Harpactea sadistica* bears needle-like structures at the end of his pedipalps, the 'arms' attached to his head which are used for passing the sperm into the female. He employs these to stab the female and deposit his sperm directly into her ovary, eliminating the need for any courtship niceties. The evolutionary explanations are not fully resolved for such behavior, but in this species such brutal tactics seem to have yielded successful results and enabled the species to thrive. In the females, some degree of shrinkage of the internal sperm-storage organs can be observed as a result, so perhaps this saves energy in growth and development and gives her a slight edge over others in the struggle for survival.

The transition from the diverse, bizarre world of invertebrates to that of the first fishes was one of the greatest major evolutionary journeys. Only recently have the ancestral roots of the first vertebrates been resolved using molecular data. The simple animals that we find on beaches called sea squirts or cunjevoi (tunicates) are now regarded as the closest living invertebrates to our kin. Although they look nothing like an early fish,

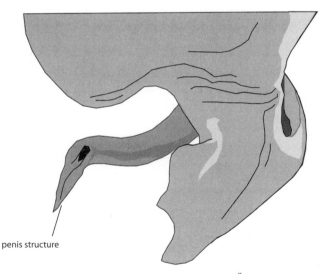

penis structure

Harpactea spider penis structure. (John Long after Řezáč 2009)

their larval forms have muscular tails, gill slits and a stiff cartilage rod that supports the axis of the body and tail. In these respects they are quite close to the first kinds of jawless fishes that appeared around 520 million years ago. Jawless fishes today, represented by the eel-like lampreys and hagfishes, all reproduce by spawning in water, the males shedding sperm over eggs.

The next big step in the evolution of vertebrates was the advent of jawed fishes, such as our lusty armor-plated placoderms. As discussed in earlier chapters, we now know that some of these had elaborate mating methods using spiky bone-covered claspers, whilst others had the smoother, more slender type of clasper that was almost certainly used as an intromittent organ. Sharks and rays alive today use similar methods of mating, although their claspers are now soft and pliable, supported internally by rubbery cartilage rods.

With this understanding of fish reproduction, the next important stage in our journey – from the origins of intimate sex in Devonian armored fishes to us – is to try to understand how fishes eventually left the water and invaded land. And, when they did so, how on earth did they have sex then?

10

Sex on the Beach

... and the grunion ran again
through the oily sea
to plant eggs on shore and be caught
by unemployed drunks
with flopping canvas hats
and no woman at all.

Charles Bukowski,
Mockingbird Wish Me Luck

If you thought Burt Lancaster kissing Deborah Kerr on the beach amongst the crashing Hawaiian waves was sexy in *From Here to Eternity*, you ain't seen nothing yet. The Californian grunion (*Leuresthes tenuis*) is a small fish up to about eight inches (20 centimeters) long that has unashamedly wild sex on the beach on moonlit summer nights. Let me paint a picture of the scene to give an idea of how it all happens.

It's about three days after the full moon and the beach is eerily lit from the pale moonlight on this warm summer night. With the washing in of the tide, a group of females make their way onto the beach. Writhing their bodies along the wet sand, they manage to bury themselves up to their arms (well, pectoral fins) with only the top part of their bodies and heads exposed. Next come the randy males, up to eight of them for each female, who surf in on the following wave and urgently wrap themselves around the exposed ladies, madly ejecting their sperm all over the females' heads and bodies. The sperm slowly washes down the females' slick skin and eventually reaches the eggs they have laid in the sand. Job done, the ladies wriggle out and head for the open water, clearly in need of a wash, leaving the fertilized eggs buried safely in the sand. The eggs will be sought out by various scavengers, like crabs and gulls, but if some survive they will hatch around ten days later with the next high tide.

Now you can perhaps understand why, in the 1950s, young Californians would go out on a summer night to see the grunion do their stuff – 'going to the grunion run' soon became a euphemism for 'going on a hot date', and Frank Zappa even captured the mood in his 1963 instrumental 'Grunion Run'.

Grunions are ray-finned fishes (or actinopterygians – fishes with bony rods or rays supporting all the fins), as are goldfish, trout or salmon, but the latter all still rely on being in water to breed. Across the 30,000 or so known species of actinopterygians there is a wide range of mating behavior, several of which we've reviewed in earlier chapters, but those that mate out of water are few and far between. Even air-breathing fishes such as mudskippers and lobe-finned lungfishes still need to mate in water, as do most of the known amphibian species, such as frogs. However it happened, at that critical time in evolution when fishes began leaving the water to invade land, such major changes in reproductive physiology as breathing air and retaining enough moisture without desiccation, and hence changes to the urogenital system, involved in mating out of water would have eventually become mandatory rather than just optional.

The first land animals were an early amphibian ancestor, the tetrapods, and were anatomically very similar to their fishy predecessors in most ways. But how was that transition from water to land first achieved? One of the most famous discoveries that helped us to answer this question was *Tiktaalik*, an advanced lobe-finned (sarcopterygian) fish. When this find was announced in 2006 it made headline news around the world and was touted as a true 'missing link' between fishes and land animals.

Discovered in Arctic Canada by a team led by Professor Neil Shubin of the University of Chicago and Dr Ted Daeschler of the Philadelphia Academy of Sciences, it took several grueling field seasons in the cold northern conditions, searching in just the right kinds of geological deposits where they predicted such a fish fossil should be, before they struck pay dirt. *Tiktaalik* (which means 'burbot', a large fish in the Inuktitut language of northern Canada) is a really neat fossil in all senses of the word. It bears a large flat head, like an alligator, with eyes on top, akin to many of the early fossil amphibians, and it has a robust shoulder girdle with stout forelimbs. But the really amazing thing about *Tiktaalik* is that it has bones on the skull and cheek that match precisely, one-for-one, with those of the earliest well-known tetrapods (four-legged animals), such as the 365-million-year-old *Acanthostega*. The paired forelimbs (or 'arms') of *Tiktaalik* internally sport a robust humerus, ulna and radius forming an arm skeleton, as does *Acanthostega*. However, the fin ends in a series of stout rays in *Tiktaalik*, denoting that it is still a 'fish', whereas *Acanthostega* limbs end in a spray of stubby digits. Recently Catherine Boisvert, now at Monash University in Australia, and Per Ahlberg, from Sweden's Uppsala University, discovered that a close relative of *Tiktaalik*, a fish called *Panderichthys*, shows the beginnings of digits forming in its fin structure, another case demonstrating the intermediate stage in the evolution of fishes to land animals.

Scientists who have studied these earliest amphibians in detail, such as Professor Jenny Clack at Cambridge University's Zoology Museum, agree that they were most likely fully aquatic animals, and that having multiple fingers or toes (*Acanthostega* has eight) was probably a specialization that aided them in swimming. Aquatic reptiles like some plesiosaurs or ichthyosaurs also had multiple rows of digits. After the first amphibians invaded land, the standard number of fingers and toes eventually stabilized at five per limb. The pattern would later be set as an asymmetric one on each limb with an uneven number of digits, as seen in our own fingers and toes. Despite such useful accessories, the first terrestrial amphibians, like nearly all amphibians today, would have returned to the water for breeding. Some of the earliest evidence of tetrapods invading land were revealed at the Zachelmie quarry in Poland in early 2010, from shoreline deposits dated at around 390 million years old. The suggestion that these creatures were probably the first ones to have our kind of sex on the beach is not too far-fetched. Whether they copulated, spawned in water or had close-up cloacal contact, as some frogs do today, remains a mystery – unfortunately no evidence of their mating behavior has been left behind in their fossil record.

Today's amphibians display a wide range of mating behaviors, but the vast majority still return to fresh water to breed in a similar fashion as ray-finned fishes (the females lay a mass of eggs and the males emit sperm over them). Like the grunion, there are many wild and eccentric variations within the amphibians, of course. Many male frogs mate by 'amplexing' on the females, a polite term for grabbing the female from behind and forcing a close cloaca-to-cloaca position. This means the sperm can either be shed directly into the females' cloaca or onto her eggs as they come out. The North American tailed frog (*Ascaphus truei*) uses a crude form of copulation that involves a penis-like appendage in the form of a modified tail. Males grab hold of the females and then insert their tail inside the female's cloaca to deposit sperm via internal fertilization, unique in the frog world. The males appear to be true gentlemen of the amphibian world, with larger, more muscular males not attempting to push off smaller males caught in the act of mating, but preferring to wait their turn.

Frogs show a wide range of mating behavior. Some, like the incredibly diverse *Eleutherodactylus* genus (rumored to contain about 700 species), are able to breed away from fresh water through a kind of direct development: young frogs hatch from eggs without going through a tadpole phase. Some even had live birth, like the rare or recently extinct golden coqui of Puerto Rico (*Eleutherodactylus jasperi*) – the eggs hatched inside the female and then the young hopped out as fully formed small frogs.

The Australian gastric-brooding frog (*Rheobatrachus*) – unsighted since 1985 and now thought to be extinct – had a unique form of parental care not seen anywhere else in the amphibian world: after the female's eggs were externally fertilized by the male she would swallow the lot, taking clutches of up to 40 large ripe eggs into her digestive system. A thick coating around the egg contained the substance proglandin, which turned off acid production in the stomach, but nonetheless a few of the eggs would certainly end up ingested as food. About half the eggs would eventually hatch and the tiny tadpoles continued to live in her stomach by secreting more of the chemical to stop digestion. As the tadpoles developed over six weeks, the mother's stomach would expand to occupy almost all of her body cavity. Eventually the tiny little froglets would emerge from the mouth of the proud mother, a sight we humans will probably nevermore behold.

Sometimes nothing will stop male frogs intent on sowing their seed. In some cases the mating frenzy of wood frogs (*Rana sylvatica*) gets so heated that the males will mount almost anything that comes along, including other males or even other species of amphibian such as unsuspecting male or female tiger salamanders. However, recent experiments have also shown that certain conditions can radically alter frog behavior. For instance, stream waters polluted by even minutely small amounts of the weed killer atrazine, now recognized as a potent endocrine disruptor, can change the sexual organs and behavior of male frogs. A paper by Tyrone Hayes from the University of California, Berkeley, and colleagues in 2010 proved that such agents can chemically castrate male frogs and induce homosexual behavior. Male frogs would mount other male frogs after being immersed in the polluted waters, and some 10 per cent would eventually change sex and become females. The findings were very significant in that it finally provided a reason for the dramatic decline of populations of amphibians, even in areas where only small amounts of chemicals seep into their breeding pools. Such chemicals are active to amphibians in measures of just 0.1 part per billion.

So, do any groups of amphibians actually copulate? The answer is yes, but they belong to a peculiarly specialized subset called caecilians.

These superficially snake-like legless amphibians spend most of their life either in water or moist tropical soils, hunting worms and other smaller prey. Some are so specialized that they lack lungs entirely, taking all their oxygen needs in directly through their skin. But they all copulate in order to reproduce, the males using a stocky little structure called a phallodeum – a penis-like structure derived from the tissue of the rear part of the cloaca – during mating, inserting it into the female and leaving it there for up to three hours whilst transferring sperm.

In some caecilians (for example, *Scolecomorphus*) the cloacal region is reinforced with cartilaginous spicules (little rods like needles), some of which also adorn the phallodeum. In a 1998 paper, Dr Marvalee Wake of University of California, Berkeley, compared these spicules with other mineralized structures that develop in intromittent organs of animals, such as the os baculum and os clitoridis in some mammals, and the mineralized parts of the hemipenes (paired penises) in some lizards and snakes. But she was quick to point out that she doesn't regard such structures as homologous or 'equivalent evolutionary derivatives' to the spicules of caecilians due to their different developmental origins. (This is a debate I will return to in the last part of this book, as recent research involving Hox genes suggests that some of these structures across a wide range of animal groups could possibly be developed by the same genetic processes, even if actual tissues don't appear to be of equivalent types.)

Frogs and caecilians are both examples of highly specialized amphibians whose body plans have evolved from the primitive long-bodied, long-tailed forms seen in most early fossil representatives. In order to get some greater appreciation of how primitive amphibians might have mated, we can also look at the habits of modern newts and salamanders.

The spotted salamander (*Ambystoma maculatum*), for example, which lives in southeastern USA and grows to about 10 inches (25 centimeters), undertakes an elaborate courtship as part of the mating ritual. The male initiates contact with a female by nudging her with his head then, after repeatedly encircling her, he deposits his package of sperm (spermatophores) on the floor of the pond for the female to pick up with her cloaca. Males are very competitive and sometimes place their sperm packages on top of other males' sperm if rivals are encountered during the mating ritual. The male California newt (*Taricha torosa*) first attracts his mate's attention by doing a little dance, then mounts the female to tenderly rub his nose under her chin, or to stroke her with his hind legs until she succumbs to his charms. It could take him up to an hour to entice her to pick up the package of sperm he has deposited on the pond floor with her cloaca. As newts and salamanders are physically closer in shape to the early fossil amphibians, we can only extrapolate that *Acanthostega* might well have had a similar kind of mating behavior.

The next stage in the evolution of land animals – from amphibians to reptiles – brought with it a big innovation in reproduction. This group of tetrapod vertebrates (animals with four legs and a backbone) are also called 'amniotes' because they have an amnion, or sac, that encases the embryo (you've seen it as the thin outer membrane on a shelled boiled egg). This can be inside a hard-shelled egg, as in reptiles and birds (and dinosaurs), or inside the animal, as in mammals. We identify early fossil amniotes (the most primitive known reptiles) by their similarities with modern reptile skull patterns. Three bones in particular – the parietal, supratemporal and tabular bones – are firmly interlocked as part of the skull roof in advanced lobe-finned fishes like *Tiktaalik* and all early tetrapods (amphibians). However, these bones become detached from the skull roof and are instead part of the occipital or back skull region in amniotes like reptiles, birds and mammals. So using these criteria we can look at skeletal fossils to determine when this stage in evolution most likely took place.

The oldest known accepted amniote fossil is that of *Casineria*, a lizard-like reptile dated at around 340 million years ago (Early Carboniferous period). With this new stage in our vertebrate evolution came the ability to lay eggs on land and thus be free from the need to go back to water to breed. The development of the amnion membrane meant that the developing animal inside the egg could still breathe while retaining moisture inside the shell. This innovation was the singular most important event in our entire evolution because it necessitated a new way of mating for this group: compulsory internal fertilization.

Today, all amniotes mate either by copulation (as in the vast majority of reptiles and mammals, plus some birds) or by close cloacal contact to transfer sperm to eggs inside the female (most birds). The eggs then develop and are either laid with a hard shell around them, as in birds and many reptiles, or developed inside the mother until ready for live birth, as in us mammals and many reptiles.

The peak of reptile dominance was the age of dinosaurs, the Mesozoic era, from around 250 to 65 million years ago. This time saw the rise of the largest land animals on Earth, the appearance of the first birds, and the origin and diversification of true mammals. By the end of the dinosaur era, reproductive biology on the planet would never be the same again.

11

Dinosaur Sex and Other Earthshaking Discoveries

Of the many strange forms of ancient animal life
brought to light by the labours of geologists and
collectors in various parts of the world, perhaps those
of the Dinosaurian order are the most wonderful.
Not only in size, but in strangeness and variety, they
may be said to stand alone. They were indeed the
veritable dragons of old time ...

Rev. H.N. Hutchinson
Creatures of Other Days

When Robert Plot, professor of chemistry at Oxford University, published a book on the natural history of Oxfordshire in 1677, he figured in it a very peculiar type of fossil found in a quarry near Chipping Norton the year before. The strange discovery could have been that of a giant's fossilized scrotum, with two large ball-like structures hung within a petrified sac, and in Richard Brookes' 1763 account of this fossil, he saw it precisely this way and named it *Scrotum humanum*. Later studies showed the testicle-like structures were really the well-rounded pair of condyles of the lower femur of a dinosaur leg, and Plot's figure of 1677 became the first record ever published of any dinosaur. English geologist William Buckland named it *Megalosaurus* in 1824.

The late, great English paleontologist Dr Beverly Halstead, who spent most of his life studying ancient jawless fishes, had more than a passing interest in dinosaur sex.

Named 'Scrotum humanum' by Richard Brookes in 1763, this fossil was later identified correctly as part of the leg bone of a *Megalosaurus*. (From Plot 1677)

Once when preparing an article about *Scrotum humanum* with fellow paleontologist Dr Alan Charig, in which they received help from two of their colleagues, Dr William Ball and Dr Robyn Cocks, he tried to include an acknowledgement in the paper to Drs Cocks and Ball. Unfortunately, Cocks was Charig's boss and didn't appreciate the humor in the acknowledgment, so the fortuitous wording ascribed above was never published.

Throughout his career, Halstead gave amusing demonstrations of possible dinosaur sex and theorized at various paleontological meetings about how they might have mated, and even published an early reconstruction of dinosaurs mounted upon each other in the act of mating in his 1975 children's book about dinosaurs. During one Canadian lecture on the evolution of humans, the audience was shocked when he showed a slide demonstrating how humans are just like other primates. 'I showed this picture of me naked in a tree with my penis hanging down,' he recalled. 'It sort of freaked out my audience.'

Halstead based most of his interpretation of their sex lives on living reptiles. He insisted that dinosaurs had only one main mating position: 'Mounting from the rear, [the male] put his forelimbs on her shoulders, lifting one hind limb across her back and twisting his tail under hers to align the cloaca.' He also mused that 'they may additionally have lain side to side, male to port and female to starboard, and sort of snuggled up together, bottom to bottom'. The golden rule, he declared, would have been that rear-mounting males must have at least one foot on the ground!

Dinosaurs were the largest animals to ever walk the Earth. Some, like the massive long-necked *Argentinosaurus*, with individual neck vertebrae some six feet (two meters) high, could grow to 80 feet (25 meters) long and weighed up to 80 tons (72.5 tonnes). So did the Earth shake for them when they mated? Maybe, but the real question here is was Halstead right? And what evidence do we have to reconstruct their sex lives?

Scouring the internet (which has become a first-base tool for scientists) for clues as to how dinosaurs might have reproduced gives mostly vague speculation. One website proposes males probably didn't have penises so must have used cloacal kissing, juxtaposing their massive backsides together for the interchange of seminal fluid to the female à la Halstead and as do most frogs and many birds. The author on this site summed up his take on the current position: '… male and female genitals don't tend to persist in the fossil record, and the average paleontologist knows less about dinosaur sex than a second-grader knows about the human variety.' The statement is largely correct on the score of fossil evidence for dinosaur reproduction, but perhaps looking at how some living reptiles and birds (now believed to be the closest living group to dinosaurs) mate might shed light on the mystery of ancient dinosaur sex.

Today on Earth we have a vast diversity of reptiles – some 8200 or so species of them. All living reptiles (except tuataras) reproduce by mating using copulation, with males having paired intromittent structures (designed to be inserted) called 'hemipenes'. In some cases this is reduced to one single organ, a penis, such as is found in tortoises, turtles and crocodiles. Crocodiles and alligators (crocodilians) are the closest living relatives to the ancestral group that branched off to form dinosaurs, birds and crocodilians about 230 million years ago. All dinosaurs belonged to the Archosauria group (meaning 'ruling reptiles'), of which the only living representatives are crocodiles.

The most primitive living reptile, though (as determined by their skull structure), is the lizard-like tuatara, a kind of living fossil that inhabits parts of New Zealand. These are true relicts – hangovers from the Triassic period 240 million years ago. The tuatara male does not have a penis but transfers sperm directly into the female's cloaca in the same way that most frogs do. It takes a long time for the tuatara to become sexually mature, about 20 years or so, and then females mate and lay eggs only once every four years. Some tuataras are true wonders for their age. In 2009, one kept in captivity for many years named Henry became a father at 111 years old, possibly for the first time, by mating with 80-year-old Mildred at the Southland Museum in Invercargill. As all other living reptiles have either a single penis or paired penises, the evolutionary logic behind all this tells us that all living reptiles most likely acquired or developed their reproductive organs independently of the ancient line containing tuataras.

The group known as squamates, or scaled reptiles, includes all living lizards and snakes, and with nearly 8000 known species they are by far the most successful of all reptiles. They have a molecular divergence time of origin estimated to be somewhere in the Late Jurassic (about 150 million years ago), well after the diversification of dinosaurs, and about the same time as the first birds appeared on the Earth. As mentioned, they have two penises (hemipenes), and to make them even more interesting these might be forked on each side, or adorned with spiky scales. This affords them a highly specialized reproductive mode that is separate from the line leading back to dinosaurs and birds, in terms of evolutionary origin. They generally insert their hemipenes into the female one at a time and alternate sides, as each side has its own sperm store.

Some of the squamates should also be admired for their monogamy and parental dedication. The shingleback or bobtail lizard of Australia (*Tiliqua rugosa*) tends to be monogamous outside the breeding season, and couples may stay together for up to 20 years. They are viviparous, and some of their young may be enormous – up to a third of the mother's body weight. Imagine a female human giving birth to a 50-pound (23-kilogram) baby!

But not all lizards and snakes share the bobtail's high moral ground. When North American garter snakes (*Thamnophis sirtalis*) mate, both sexes emit pheromones that enable other snakes to smell whether they are males or females to help them home in on a partner. But the coolest snakes, so to speak, are the warmest males – these always win the females. Up to 25 randy males may cluster around a single female, forming an orgiastic ball of writing, copulating snakes. Sometimes male garter snakes will emit female-like pheromones to fool other males into an attempt to mate with the 'she-male' snakes; scientists think this behavior tactic is designed to help them get warm quickly after their winter hibernation. The warmed-up 'she-males' then have a more successful time mating with the female snakes than their decidedly cooler competition. In order to test this hypothesis, a team led by Dr Rick Shine from the University of Sydney fixed miniature thermal data loggers to snakes to accurately measure their heat transfers, and also used dead snakes as courtship targets. They proved their point and published the result in *Nature* magazine. And yes, some randy snakes did try to mate with the dead snakes – even necrophilia is not out of bounds in the animal kingdom, as another case involving tortoises (described below) demonstrates.

As said, male tortoises and turtles mate by using a single organ, a penis, albeit often quite slowly. Giant Galápagos tortoises can live for up to 150 years and grow to over 50 pounds (23 kilograms) in weight. When these giants mate, they don't do anything in a hurry. After bobbing up and down to attract the female's attention, the male charges and rams her shell, sometimes nipping at her legs to make her crouch back into her shell. Then the male mounts the female from behind and spends a good few hours copulating with her (and tortoises have some of the most gruesome male genitalia I've ever seen adorning the glossy sealed pages of zoological mags). Funnily enough, the male tortoise penis is similar in many ways to that of a mammal's, and some scientists have reported that the two are functionally very close in hydrostatic structure, that is, in order to become erect they use a similar array of collagen fibers. Even more surprisingly, given their slow and dull reputations, observations made in 1971 on the mating behavior of giant Aldabran tortoises (*Geochelone gigantea*) casually noted that one male tried to mount another dead tortoise while another fed upon the dead carcass.

Nearly all reptiles with tails mate by copulation and

The penis of a tortoise (right) is similar in many ways to that of a mammal, whereas other reptiles like snakes and lizards have paired hemipenes. (From Jones 1915)

rely on a penis to bridge the gap left when a male mounts a female and tails get in the way. In some cases the only evolutionary option is to enlarge the penis so it can reach further and do its job, hence the very large penises of tortoises compared to all other reptiles.

So now that we've looked at some living reptiles, do any fossil reptiles related to these tortoises and squamates preserve clues as to their reproductive behavior? Large marine reptiles called ichthyosaurs have been found with remarkably well-preserved embryos, some even caught in the act of childbirth, indicating that it's most likely they were having copulatory sex in a similar way to dolphins. It's easy to deduce this as the idea that they used cloacal kissing to exchange gametes is absurd when you consider their inflexible, fish-like body shapes. These were wholly aquatic creatures, not adapted to crawling onto beaches, so sex must have taken place in the open water much like dolphins or large sharks do today, and both of these groups engage in copulation.

Another group of small marine reptiles known from sites dating back to the Triassic of China, including forms like *Keichousaurus,* also gave birth to live young and have been found with embryos inside them. This group was ancestral to the large long- and short-necked plesiosaurs, some of which grew up to 50 feet (15 meters) in length. Once committed to the strategy of live bearing, it was most likely that this mode of reproduction carried through to the whole group. A specimen of a plesiosaur called *Polycotylus* at the Natural History Museum of Los Angeles County has recently been prepared for display in a new exhibition. During the process it was discovered by Drs Luis Chiappe and Robyn O'Keefe to have an embryo preserved inside it, which confirms for the first time that plesiosaurs had internal fertilization, and thus copulation was their most likely method of mating.

Mosasaurs were marine-living monitor-like reptiles closely allied to today's goannas of Australia except they grew to fearsome sizes, up to 50 feet (15 meters) long. They too were viviparous. A specimen of a small animal, *Carsosaurus* from the Cretaceous of Slovenia, was found to have at least four advanced embryos inside it, indicating that these creatures probably gave birth to their young in the water (which would have been tail first).

All of these examples suggest that we can assume, from these cases of live bearing in streamlined aquatic fossil reptiles, that they most likely mated by copulation and, like whales today, the penis (or hemipenes, perhaps) would have been stored inside the male's cloacal region to emerge for mating purposes only, and not upset the smooth, streamlined shapes of their aquatic bodies by creating drag. After all, we don't see whales swimming along with their penises hanging out unless they mean business.

An interesting fact about the mosasaur's closest living relatives, the monitor lizard, is that the largest species of the group, the komodo dragon (*Varanus komodoenis*), was shown in 2006 to be able to reproduce asexually by parthogenesis. Two females held in European zoos without access to males laid fertile clutches of eggs which, when the DNA was analyzed, showed they were perfect clones of the mother. This was the largest (in terms of size) recorded case of parthogenesis happening in vertebrates, and at the highest evolutionary stage (that is on the scale from low, fishes, to high, humans). Perhaps, in the absence of Mr Right, some female mosasaurs had this latent ability too.

Some years ago, my good friend science journalist Carmelo Amalfi and I got together to tackle the issue of how dinosaurs might have done the deed, and what kind of gear they might have used to do it. In 2005 he interviewed a number of leading paleontologists and wrote a very comprehensive article on the subject. Many of the scientists he spoke to believed that dinosaurs must have mated via cloacal kissing rather than via males bearing large penis-like organs.

The only evidence that has so far come to light to support the view that dinosaurs were sexually dimorphic – that males were of differing size or shape from the females – comes from observations of some *Tyrannosaurus* skeletons as being gracile and others more robust (presumably the female). This data is not universally accepted by all paleontologists as the sample size is very small, however some other data on the common occurrence of tail vertebrae that fuse together in the large, long-necked dinosaurs *Apatosaurus*, *Diplodocus* and *Camarasaurus* may back this theory up. In a 1991 paper, authors Bruce Rothschild and David Berman concluded that perhaps 'the fused caudal vertebrae occurred only in the females and represent an adaptation permitting the upward and sideways arching of the tail to facilitate copulation'. Since the paper was published, other paleontologists have criticized this interpretation saying there is not enough statistical evidence to prove a case for sexual dimorphism in these samples, so the debate still continues.

The only other useful piece of fossil evidence

to impact on our understanding of how dinosaurs reproduced was the discovery of a fossil dinosaur pelvis region from the oviraptorosaur group, with two eggs preserved in place in its oviduct. The specimen came from the Jiangxi Province in China and was dated at around 70 million years old. In describing the eggs, lead author Dr Tamaki Sato, from the Canadian Museum of Nature, stated that the arrangement of the eggs indicated that this dinosaur retained two functional oviducts like crocodiles do, but had reduced the number of eggs ovulated to one per oviduct, as in birds. Thus we see a crocodilian style reproductive tract with an avian style of egg formation. To me, this supports the possibility of such dinosaurs having a crocodilian style of reproduction (using a penis for copulation) and laying eggs like a primitive bird such as ostriches and other ratites (a group of mostly large flightless birds) would, two at a time – as opposed to laying eggs randomly (in no neat order), as would reptiles such as turtles or crocodiles.

Various experts on biomechanics have weighed in to the debate, claiming that it would have been unsafe for animals of such massive weight to copulate, although estimates of dinosaur weight and biomechanical limits have been widely criticized as being imprecise. One piece of evidence that can't be ignored is the mathematical calculations based on blood pressure by Professor Roger Seymour of the University of Adelaide. His study of the giraffe showed that its blood pressure is roughly twice that of other mammals, and its heart needs be proportionately 75 per cent larger due to the physiological constraints of its long neck and highly perched head. Bearing this in mind, he suggested that long-necked dinosaurs could only have mated in a particular way:

> If you calculate the vertical distance between where the heart is and where the head is – and some of these animals grew to up to 30 feet [nine meters] – you can measure accurately what the minimum pressure is at the bottom of a 30 foot column of blood. It works out to be seven times the normal mammalian blood pressure. Rear mounting is not a big problem if one keeps the neck horizontal.

Just imagine a 70-ton (65-tonne) giant sauropod fainting after loss of blood pressure to the head at the time of orgasm while mounting its mate. Yes, the Earth certainly would have shaken for them.

In recent years, scientists have determined that all

living birds are descendants of the mighty meat-eating theropod group of dinosaurs (think *Tyrannosaurus rex, Allosaurus*). The evidence for this has come mostly from a series of remarkable fossils found in Liaoning Province in China since the mid 1990s, showing well-preserved dinosaurs with feathers covering their bodies. And these weren't just simple feathers – some have quite complex branching and wispy down, of the kind only otherwise seen in living birds. To understand this link, first we need to set the scene by looking at reptiles and birds in terms of their general anatomy.

Good early anatomists like Dr Thomas Henry Huxley (aka 'Darwin's Bulldog'), who studied the first bird fossil, the late Jurassic *Archaeopteryx*, back in the late 1860s, immediately recognized the skeletal similarities between the early fossil birds with teeth and long bony tails to small predatory dinosaurs like *Compsognathus*. Since then, thousands of well-preserved dinosaur finds have really nailed the connection that birds are their descendants. All stages in between the two are clearly represented by actual fossils: from small running dinosaurs with simple feathers (such as *Sinosauropteryx*), to larger predatory dinosaurs with elaborate feathers (like *Caudipteryx* and *Anchiornis*), to dinosaurs with feathered wings that could glide (*Microraptor*), to early birds like *Archaeopteryx*.

Living birds, represented by some 10,000 species worldwide, are the most diverse group of land-living vertebrates. As far as reproduction goes, most male birds lack penises, except in some of the primitive groups like ratites (which includes ostriches and other large flightless birds), and in certain ground birds like ducks and geese. Having already encountered the spectacularly long male organ of the Argentine duck in Chapter 1, it might come as a surprise to learn that the largest living bird, the African ostrich (*Struthio camelis*), has a very small penis, only about eight inches (20 centimeters) in length for a bird up to eight feet (2.5 meters) tall and weighing over 220 pounds (100 kilograms). Ostriches, which are widely farmed for their meat, eggs and feathers, have been observed mating both in the wild and under constrained conditions. In the wild males display an elaborate courting dance, flapping their wings against their bodies and whacking their heads from side to side to dazzle the female. One scientific paper describing the male's delightful wooing strategy says: 'At the time that they work up to full sexual display, the cloaca and penis turn bright red. Having reached this level of excitation, the male may urinate, defecate and display its erected penis.'

This seemingly works well for female ostriches. Then again, perhaps it explains why some ostriches have turned gay – female ostriches in captivity have been observed mounting other females. Other examples of avian homosexuality include the two chinstrap penguins at New York's Central Park Zoo, Silo and Roy. After entwining necks, calling to each other and going through the mating motions, they built a nest together. When an egg from another nest was put into theirs, Roy and Silo took turns warming the egg until it eventually hatched into a young female which was named Tango. Despite a six-year partnership, Roy and Silo eventually split ways when a cute female penguin named Scrappy caught Silo's eye and he left Roy for her. The odd twist to all this is that Roy and Silo's 'adopted' daughter Tango eventually chose another female mate to pair up with.

But as far as living birds can contribute to our understanding of dinosaur reproduction, most of them – that is, the 'cloacal kissers' – are far too specialized in their methods to be informative. The vast majority of living birds are passerines, or perching birds, which are not penis bearers but mate by transferring sperm from the male cloaca to the female. Some have refined this art to an amazing degree, such as the dunnock, which can mate in flight in less than a second!

A recent molecular study concluded that ostriches and other primitive flightless birds (collectively called 'paleognaths' because of their archaic palate structure) are indeed the most primitive members of the living bird group, with ducks, geese and some other waterbirds next up the evolutionary ladder leading to the crown group containing all the rest of the birds.

Taking an evolutionary perspective to this scenario, it would seem that most modern flying birds have lost the penis secondarily; that is, after it was first widespread across the group as a whole. Perhaps they did this to save weight for flight, or simply in accord with the necessary anatomical modifications to the tail and cloacal area that came with the ability to fly. This enabled simple cloacal kissing to be more effective than copulation and all the physiological modifications to anatomy that goes with it. Despite this loss of penis, many flying birds have evolved elaborate and beautiful courtship rituals, such as the bird of paradise of New Guinea and the Australian bowerbird. Primitive, ancient birds probably did possess a penis, so it's quite likely their previously closest ancestors, the meat-eating theropod dinosaurs like *Tyrannosaurus*, also mated using an eversible penis (projecting from inside the body to outside by means of the tissue literally turning inside out), most likely a terrifyingly large one. For an animal the size of *Tyrannosaurus* (40 feet, or 12 meters, long) to mate effectively it would need a penis in the order of at least six feet (two meters) long, maybe a lot more if it happened to be corkscrew-shaped like a duck's.

So the most likely case for dinosaurs, following what we know of their living descendants, primitive birds, and their closest living relatives, crocodilians, is that they mated using copulation. For very large dinosaurs weighing tens of tons, as do elephants today, it must have been a somewhat delicate business, balancing vast weight on the female's back and certainly involving male reproductive organs that evolved on a truly grand scale, perhaps matching or exceeding those of the living whales.

It seems not unlikely that one day paleontologists will find a fossilized dinosaur penis. Recent soft-tissue fossil finds are coming to light as new sites are discovered and greater detail can be uncovered using new technologies, such a studying fossils with infra-red photography, X-rays and other methods of tomography. Recently, some of our 380-million-year-old Gogo fish fossils were found to have individual muscle cells with nerve cells still attached to them, and the umbilical cord of the *Materpiscis* fossil is another such example of remarkable soft-tissue preservation. A small dinosaur fossil found in 1981 in central Italy, named *Scipionyx*, also showed excellent soft-tissue preservation, with clear impressions of the intestines, liver and some of the muscles. So the day will surely come when a remarkable new dinosaur fossil pops up solving the age-old mystery of how dinosaurs really did the deed.

12

We Ain't Nothing but Mammals

It also teaches us that such oversized structures
[penises] evolve in several alternative ways that
biologists are still struggling to understand. Thus even
the most familiar and seemingly most transparent
piece of human sexual equipment surprises us with
unsolved evolutionary questions.

Jared Diamond,
Why Is Sex Fun?

Rated number 49 in one survey of the 100 worst songs of all time, the lyrics to the Bloodhound Gang's song 'The Bad Touch' nonetheless contain a quintessential truth: we may be nothing but mammals, but that isn't all that bad. Indeed, there are even some redeeming features of our specialized anatomy. So for all those mammals reading out there, here is the chapter you've been waiting for: a quick recap on the ins and outs, so to speak, of mammalian sex.

Mammal reproduction relies upon a penis on the male to transfer sperm into the vagina of the female, so the eggs are fertilized internally and nurtured there until the foetus is developed. Some primitive mammals like monotremes, including the egg-laying platypus and echidna of Australia and New Guinea, lay leathery eggs, but they are the real exception to the rule. Marsupials, like most of the Australian mammals (kangaroos, koalas, wombats and so on), give birth to a tiny foetus which crawls up into its mother's pouch and is there suckled until it's large enough to leave the pouch and fend for itself.

Placental mammals, the vast majority on the planet, such as primates (humans and apes), dogs, cats, whales and mice, all develop the unborn young inside the mother, fed by a placenta until it is at a stage of advanced development where many are born almost independent. We primates are the exception to the independence rule in that our babies require parental care for much longer before they can look after themselves, and this is due in part to the rapid increase in our brain size in recent evolution.

But in order to comprehend what being a mammal is really all about, and how our sexuality evolved from those reptilian ancestors, we need to take a journey back to the verdant forests of the late Triassic period, about 220 million years ago, when mammals first appeared on Earth.

Reptiles evolved into mammals almost seamlessly, as shown by the superb fossil record preserving the many intermediate stages between archaic reptiles to mammal-like reptiles to true mammals. Looking at skeletal remains today, we principally define mammals through one main character: having a single bone forming the lower jaw (the dentary). The other small bones present on the jaw in their ancestors, the reptiles, became incorporated into the inner ear in mammals (such as the incus and malleus). This stage of evolution occurred about 200 million years ago when shrew-like forms such as *Hadroconium* appeared with a true mammalian jaw joint, even though its inner ear was not quite fully mammalian at that stage.

As mentioned, all male mammals have a penis used primarily for mating, but the shape, relative size and use of the penis, as well as its internal structure, differs widely among the group. Three main kinds of mammalian penis have evolved. The first is made of fibroelastic tissue, as found in cattle, pigs and whales. These are always semi-erect and the tissue, being dense and firm, enables the penis to be expanded in thickness but does not increase its length by much, so it's better adapted for staying in during copulation to ensure longer times for sperm transfer. Pigs, for example, have coiled penises which can lock inside the female's vagina and reach the uterus for extended periods of sloppy swine sex.

The second kind of mammalian penis is that supported by a bone called an 'os baculum' (or 'os priapri'). The baculum (as it will be referred to from here on) exists as a single large ossified unit inside the penis of carnivorans (dogs, seals, bears, weasels and their kin), rodents (but not rabbits), bats, insectivores (including moles, shrews and hedgehogs), and most primates, but not in us humans. There is always an exception and sometimes, as in extremely rare pathological cases, some human males have developed a penis bone. The 'os clitoridis', also called the baubellum or clitoral bone, is not as well developed in females, but occurs in many of the same groups that have the baculum.

The third kind of mammal penis is the vascular penis, exemplified by us humans. It has a very spongy structure, enabling it to increase in size dramatically from a flaccid resting state when blood fills the tissue to erect it. A cross-section of a human penis (ouch) would show a large corpus cavernosum, a sponge-like high-pressure system responsible for increasing penile size and stiffness, a smaller area of corpus spongiosum (which protects the urethra that delivers the seminal fluids or passes urine) and a muscular area. The walls of the corpus cavernosum contain layers of highly organized collagen fibers which straighten out during erection, making the penis larger and more difficult to bend.

Mammalian penises vary enormously in shape, size and uses. Robert Todd's 1852 *Cyclopedia of Anatomy and Physiology* described the extraordinary shapes of rodent penises as follows:

> In most of the Rodentia the penis contains a
> bone, imbedded in the substance of the corpus
> cavernosum. But the most remarkable part of the
> penis in the order before us, is the glans, which
> in many species is armed with such a formidable
> apparatus of spines, saws and horny spikes, that it

must indeed be a rather stimulating instrument of excitement.

Todd's illustration of the penis of an agouti, a South American rodent related to guinea pigs, presents the gruesome image of a spike-covered head with two sharp saw-edged points protruding. A modern description of the male agouti penis also describe it as 'U-shaped', an image which is slightly alarming and for which I have no definite explanation, but we could take a guess.

Some mammals develop penises with such savage accessories mainly for the purpose of cleaning out the copulatory plugs formed by the males of previous matings. The male guinea pig, a close relative of the agouti, produces a one-inch (three-centimeter) long copulatory plug that can weigh as much as half an ounce

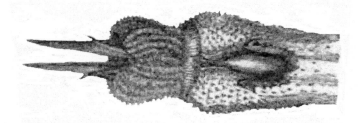

The penis of the agouti rodent epitomizes the Swiss-army-knife approach to mating. It uses its accessory spines to cut through copulatory plugs. (From Todd 1852)

(1.5 grams) and which it deposits in the female, so when another male comes along, like the agouti, its penis must either penetrate though this plug or try to remove it.

Indeed, rodents have such variability in their penis shape and accessory penile adornment that the penis alone has been used as the focus of a study to determine the evolutionary history of the rodents of New Guinea. Research undertaken in the sixties by Dr William Lidicker on some 28 species involved a close study of the 66 different characters of 72 rodent penises, and I'm told by a mammal expert that his classification of the rodents by and large holds true today.

But why did evolution develop so many weird designs for reproductive organs? Bill Eberhard of the University of Costa Rica has suggested the male genital organs evolved specific shapes mainly to stimulate the female during copulation so she could use this as a method to assess the quality of his sperm. This is contra to the previous 'lock and key' hypothesis, which held that individual shapes evolved specifically for individual species so as to prevent other species cross-mating, an idea which is no longer favored by most biologists. Genital diversity is now explained by evolutionary biologists according to four major hypotheses, but common to them all is that it is either driven by female choice for mates or sperm competition (as we shall see in the next chapter).

But back to mammals, and how we came to be as we are. With three kinds of penis known, and understanding from modern observations that all mammals mate by copulation, what can we deduce about the origin of mammalian mating systems? Can fossils shed any light on early mammalian copulatory behavior?

In fossil mammals we finally do see real evidence of sexual behavior because of the ossified penis bone of many mammals. A 12,000-year-old fossil baculum (penis bone) of a walrus measuring five feet (1.5 meters) long sold at a 2010 auction for US$8000 to a Ripley's *Believe It or Not!* museum. In Alaska, modern walrus bacula are often polished and sold to tourists as knife handles called 'oosiks'. Fossil bacula date back as far as 49 million years ago, with the early primates *Godinotia* and *Europolemur*, from the Eocene of Germany, having bacula well preserved (and often very large).

Correlating baculum length with observed mating behavior in primates demonstrates that longer bacula (implying longer penises) mean longer intromission (or copulation time), as described from work done by

Dr Alan Dixson of the Centre for Reproductive Biology in Edinburgh. In essence, though, the presence of bacula in fossil mammals doesn't really tell us anything more than we could probably guess from their presence in their living counterparts, other than confirming that they probably mated in much the same physical way. If that's the case, what have we recently found out about mating behaviors and sexual diversity in living mammals that can perhaps help fill in the gaps in our knowledge of how fossil mammals mated?

Let's start with the most primitive living mammals today. Monotremes, or egg-laying mammals, are known from two lines of diversely different living representatives, the platypus and the echidna. For a long time we knew very little about their mating behaviors and only recently has the strange sex life of the short-beaked Australian spiny echidna (*Tachyglossus aculeatus*) been revealed.

Echidnas, like porcupines and hedgehogs, have many sharp spikes over their backs, and the answer to the old joke about how do they mate is 'very carefully'. Researchers at the University of Tasmania, led by Dr Gemma Morrow and Dr Stuart Nicol, recently concluded a three-year study of echidnas using ultrasound and hidden cameras in their burrows. The team's surprising discovery was that echidnas have group sex, with one female mating with up to five males in the burrow, which added to earlier research revealing that males coming out of hibernation early sometimes try to mate with sleeping females. However they do it, echidnas are indeed successful as they are a widespread species across Australia.

The echidna's close cousin, the duck-billed platypus (*Ornithorhynchus anatinus*), has not been quite so fortunate and since European colonization is now restricted to the rivers and streams in eastern Australia; it is extinct in South Australia. Little is known about its mating in the wild, but in captivity it has been observed that males nuzzle the females and sometimes hold their tails with their bills when they get amorous. Whether the sharp, venomous ankle spines of the male platypus are used when mating with females in some way, or used in battles with rival males, remains a mystery. We know that platypus-like fossils such as *Teinolophus* go back at least 120 million years in Australia, so whatever it is these private little mammals are doing to procreate, they appear to have been doing it right for a long, long time.

In recent years, some sexual behaviors known in humans that do not directly lead to procreation have also been observed in other mammals in the wild. Such behaviors can play a secondary role in mating, whether by strengthening relationships within a group, partner bonding or stimulating arousal before the sexual act to maximize mating success. A 2008 review on the subject of animal homosexuality indicated that about 1500 species of animals, from insects through to humans, have now been observed to exhibit some form of homosexual or bisexual behavior. In some instances behavior of this kind is initiated by stressful conditions, as seen in a case of caged Australian koalas (*Phascolarctos cinereus*).

A team at the University of Queensland observed female koalas living at Brisbane's Lone Pine Koala Sanctuary apparently trying to mate with each other, with up to five individual females huddling together. Female koalas even took on the role of bellowing out male-like mating calls. Some 43 instances of this kind of behavior were observed by the researchers in what they call pseudo mating and attribute largely to stress caused by enclosure. The behaviour is conducted only between females in oestrus, as a physical sign they are ready to breed with a male. Clive Phillips, one of the team, believes that 'homosexual' behavior in some animals preserves sexual function under stress; he further points out that in the wild, koalas, which are strictly solitary animals, are all heterosexual.

Performance in mating varies enormously in mammals. The biker motto of 'live fast, love hard, die young' is particularly suited to the mouse-sized Australian marsupial predator *Antechinus*, of which there are ten species. Like some other mammals this group is 'semelparous', meaning that they mate once in their lifetime and die soon afterwards. The male *Antechinus* will mate in winter, stripping off excess weight and protein to make itself a lean, mean mating machine. It then mates with the female non-stop for up to 12 hours, after which its immune system is shot to pieces and it goes off to eventually die. The mating frenzy can sometimes turn into something of an orgy, with several males mating with the same female. She stores the gathered sperm for up to three days before fertilizing her eggs.

Moving on to the sexual preferences of primitive placental mammals, such as bats and shrews, there is again not much evidence in the fossil record. But turning once more to living examples, we find some extraordinary, and quite unexpected, sexual behavior in a particular species of Chinese fruit bat. In 2009, Dr Min Tan and his colleagues from universities in China and England announced to the world their discovery that the short-nosed Chinese fruit bat (*Cynopterus sphinx*) regularly practices fellatio. By performing this act it increases the length of copulation time, thus contributing to greater reproductive success. An excerpt from their summary explains:

> Female bats often lick their mate's penis during dorsoventral copulation. The female lowers her head to lick the shaft or the base of the male's penis but does not lick the glans penis which has already penetrated the vagina. Males never withdrew their penis when it was licked by the mating partner. A positive relationship exists between the length of time that the female licked the male's penis during copulation and the duration of copulation. Furthermore, mating pairs spent significantly more time in copulation if the female licked her mate's penis than if fellatio was absent.

Most notable is the observation that females licked the base of the male penis *during* copulation to help prolong the act, and that males licked their own penises for several seconds following the act. This was the first time, outside of humans, that any mammal has been recorded regularly practicing this act as part of the stimulation leading to mating, more or less as foreplay. So it seems that we are not the only ones in the animal kingdom who engage in highly varied sexual practices, yet it still holds true that we humans are the only mammals to enjoy using a wide repertoire of foreplay and copulatory behaviors on a regular basis to enhance our sexual pleasure.

Asian fruit bats sure are a weird lot. In 1994 Dr Charles Francis and his team reported in the journal *Nature* that male Dyak fruit bats of Malaysia spontaneously lactate, helping feed the young alongside the lactating females. This was the first time spontaneous male lactation was observed in a wild animal population, and reminds us of the evolutionary reason why males still bear teats, that although they are mostly an evolutionary 'leftover', they can potentially become active again if survival pressures become strong enough.

Most people who keep cats are familiar with the stink of tomcats spraying their scent around the place. They do this to both keep track of other males invading their territory and to advertise themselves to any females who enter the territory. Cats, large and small, are polyestrous, that is, the female can mate at any time and do so several times a year. When lions mate they may couple for up to two days and enjoy some 20 to 80 copulations each day. Cats, including lions and our smaller domestic pets, are another species that exhibit a particularly brutal form of copulation, as the male cat has a penis adorned with many short, sharp, rearward-facing spikes that rake the female's vagina as he withdraws himself. This is why female cats often give a loud cry when the male dismounts. The act has twofold purpose: firstly to remove any sperm left from a previous mating, and secondly, and more importantly, to stimulate ovulation in the female cat.

Dogs and their kin have their own peculiarities when it comes to mating, particularly in some of the more extreme artificially bred varieties. All dogs are descended from wolves, and domestication has been underway for the past 15,000 years or so, since dogs began cohabiting with humans who selected and bred them for their personality traits. This process concentrated their genes in those breeds which get along best with us. In recent years, most of the crossbreeding has been for specific needs, such as hunting or retrieving, and even more recently for novel traits that appeal to certain pet lovers. This has resulted in some real genetic oddities; male French bulldogs, for example, are now incapable of mounting the female, so artificial insemination is the only way they can propagate.

But the most bizarre sexuality of all the dog-like mammals is certainly found in the bone-crushing spotted hyenas of Africa (*Crocuta crocuta*). Female spotted hyenas have evolved a clitoris which, supported by a robust internal clitoral bone, forms a penis-like structure called a peniform clitoris, which can become erect and protrude up to seven inches (18 centimeters) from her body. For centuries this phenomenon gave rise to legends of the female spotted hyenas being hermaphrodites, but the truth of their reproductive cycle is even more strange than that: these hyenas are the only mammals in which females mate and give birth through a small slit in the female's clitoris, rather than via the regular vaginal tube.

A recent study of their anatomy by Gerald Cunha and colleagues at the University of California at San Francisco noted that the hyena pups are born through a 'tortuous' recurved passageway, eventually emerging through a small hole in the meatus or opening of the clitoris. To mate, the male has an extremely small target to aim for, and the penis uses a special hinge for flipping it during copulation which helps him get exactly the correct angle for entry.

So why would a female hyena develop a penis-like clitoris? Apparently it could come down to specific behaviors that hyenas have evolved to regulate their social structure, such as mock copulations, and to facilitate mating. As Cunha and his colleagues note, hyenas regularly inspect each other's genitalia, male and female alike, on meeting: 'Hyena etiquette requires that the subordinate hyena initiate the ceremony by presenting its erect genitalia for inspection by the dominant animal.' Female hyenas also use their erect clitoris to signal their rank amongst other females in the pack. And the small clitoral opening of the female hyena gives her more say in when she wants to mate and with whom. So despite the common occurrence of severe damage to her clitoris, which is often ripped apart by the birth of her pups, the evolutionary advantages in controlling the mating act, and selecting the best mates, seems to win the day.

It's clear that modern mammals show the wildest variations in sexual behavior of any of the higher animal groups, with us humans having the largest menu of all in the cafe of sexual preference. There may be a 'standard' way of mating to achieve reproduction (that is, missionary-style coitus), but many humans have developed tastes for mating behavior that is not solely intent on reproduction but for sexual pleasure alone.

In his book *Why Is Sex Fun?*, Jared Diamond points out

that humans are one of the few groups of mammals where females do not visibly advertise that they are fertile and in oestrus (ovulating). We have seemingly developed our sexuality around the male wanting a female practically all the time, as our females don't display visible signs of being on heat as do other primates. The theory goes that early on in the evolution of our culture, trading sexual favors for the long-term benefits associated with a monogamous relationship (mainly protection and cooperative sourcing of food) was a prime factor in shaping modern mating behavior. While many books have been written about human sexuality and how it might have evolved, my aim is to now try to frame this extraordinary change within a broader evolutionary context.

In our transition from ancestral Australopithecine ape-like hominins, living 2.5 to 3 million years ago in Africa, to modern humans, primate brain size and complexity increased rapidly, doubling in size in a relatively short time. Indeed from birth to adulthood, the human brain expands by a factor of 3.3, compared with 2.5 in chimpanzees. The reason for this brain immaturity at birth is that infants with large heads, and of course their mothers, would be at severe risk during childbirth because there is greater chance of prolonged or obstructed labor, causing oxygen deficiency and possibly death to the infant. Thus infants with small heads and immature brains at birth were favored through natural selection. But while the human brain has increased in size dramatically relative to overall body mass during the past two million years of our evolution, female pelvis size has not increased as much proportionately. This explains why we modern humans have so much trouble in childbirth relative to other mammal species.

Neanderthals, a related species of human living until about 30,000 years ago in northern Europe and Asia, actually possessed larger brains than modern humans. Newborn infant remains found at cave sites in Russia and Syria indicate that their babies were born with similarly large brain sizes as modern humans. Neanderthal life history was similarly slow in terms of development time for babies, or even slower-paced, than for us modern humans. The study of DNA by Cory McLean and colleagues has determined that we humans have lost certain genes that other mammals, like chimps and mice, still have (termed 'junk DNA'). The loss of such genes as AR and GADD45G could account for why we are more 'human' than other primates. The AR gene which develops spines (or horny papillae) on the penis of mice and chimps is lost from the human genome, hence giving us smooth spineless penises. A study by Professor Svante Pääbo and colleagues at the Max Planck Institute in Germany sequencing Neanderthal DNA has determined that Neanderthals also lost those genes. Thus, however brutish your image of the archetypical caveman is, we know now that at least they had modern smooth penises, and this implies that they took more time in copulation, and possibly enjoyed it more, than other primates. If that's not a prime factor of making us modern humans and separating us from other animals, then what is?

To understand our own sexuality as advanced primates, we need to look at other living primates to see what is the so-called norm and what might be unusual, or specialized, sexual behavior. Of the many studies undertaken of primate sexuality, perhaps the most notable examples of widely varying sexual behavior are found within the bonobo chimpanzee population.

Bonobo chimpanzees (*Pan paniscus*) are a smaller species than the regular chimp (*Pan troglodytes*) which lives in the central Congo region of Africa and holds the honor of being our closest living relative (we share 98 per cent of our DNA with them). As far as their mating habits go, bonobos are somewhat unconventional compared with other primates. Unlike gorillas and regular chimpanzees, which commonly mate like dogs, with the female's back facing the male's front, most bonobo copulations are typically missionary style as with humans. Furthermore, the female is at most times sexually active and willing to copulate, rather than just during a short period of oestrus. Adult females commonly touch their genitals against each other (called genital-genital or GG rubbing), making loud squeals suggestive of attaining orgasm. Males may use a pseudocopulation act where one male rubs his scrotum against the buttocks of another, and at times they engage in 'penis fencing', in which two males hang from a branch and rub their erect penises together. Such acts might be useful as a way of strengthening the bonds within the group. In Zaire's Lomako Forest, Nancy Thompson Handler observed that bonobos would engage in sexual activity after entering a ripe fig tree loaded with food, then sit down to eat afterwards; it has been suggested that their excitement over food might translate into sexual arousal. Bonobos also engage in fellatio between juvenile males and between males, and females, primarily for play purposes.

Other forms of homosexual behavior were documented in the 1970s for stump-tailed monkeys (*Macaca arctoides*) by Dr Susan Chevalier-Skolnikoff of the University of

California, San Francisco. The monkeys were a captive colony and some 36 incidents of homosexual behavior were observed: 'Male homosexual encounters included prolonged manual genital stimulation, sometimes mutual, oral genital stimulation (fellatio), also sometimes mutual; dorsal mountings with pelvic thrusts and, occasionally, anal intromission.' After observing other colonies of captive monkeys, Dr Chevalier-Skolnikoff noted that these particular monkeys seemed to have 'a propensity for such behavior'. The conclusions in the paper also noted that such homosexual behavior in the monkeys often followed, and appeared to be elicited by, viewing heterosexual activities. The acts were considered to be pleasurable or 'erotic', in the human sense, and when performed by juvenile monkeys were interpreted as potential training for adult sexual roles.

I asked a colleague of mine, Dr Aldo Poiani, author of a recent book on animal homosexuality, for his opinion of what homosexuality means to our evolution in general and his reply goes a long way to explaining these kinds of behaviors:

Biologists regard homosexual behavior, also known as same-sex sexual behavior, as an evolutionary puzzle, a puzzle that can be resolved by careful study of the sexual patterns shown by animals, humans included. The two major striking patterns of same-sex sexual behavior are that it is usually displayed by individuals who also mate with members of the other sex; such a pattern can be described as bisexuality, but in some less frequent cases individuals have been found who only mate with members of the same sex. The latter are known as exclusive homosexuals and they have been described in male sheep, male cattle and, of course, male and female humans.

Resolving the evolutionary puzzle of bisexuality is relatively easy, first because bisexuals do not necessarily reproduce less than heterosexuals, and second because same-sex sexual behavior can fulfill adaptive functions as mediator of dominance or cooperation. In social species, dominants may reinforce their status by mounting subordinates. This has been described in many social mammals and also social birds.

But what about exclusive homosexuality? How can evolution produce members of a sexual species that actually only mate with individuals of the same sex? For a long time the traditional view has been that homosexuals are just a product of a 'mistake', some kind of malfunction of the evolved mechanisms that produce adult heterosexuals. More recent studies, however, have supported a different view, one that sees homosexuality as an adaptation in certain social and demographic environments. For instance, male exclusive homosexuals among humans tend to be more common in large families, they tend to be younger siblings, have some tendency to be cooperative, and their relative frequency in the population is low (about 3%) and roughly constant across many diverse cultures. These patterns could be explained, for instance, if recurrent mutations encoding for the homosexual trait are selected thanks to the benefits that homosexuality confer to the homosexual's relatives, such as helping them, a process known as *kin selection*, and the association between homosexuality in some male offspring and high reproductive rates of the mother, a process known as *sexually antagonistic selection*.

This adaptive scenario of exclusive homosexuality seems to be puzzling in itself simply because if homosexuality is so beneficial, why on earth are homosexuals so discriminated against, especially in western cultures? Well, perhaps it is the relatively recent spread of negative discrimination against homosexuals in western cultures that is maladaptive, not homosexuality as such.

Indeed, autoeroticism or masturbation has also been widely observed in the animal kingdom. It is particularly well known in various species of apes and monkeys (just go to any zoo and see for yourself). Horse breeders know that it is common for ungelded males to rub themselves up against things to seek sexual release. Female porcupines have been observed using a stick as a dildo to pleasure themselves. And we have probably all experienced dogs that mount our legs – another case in point.

The great sexologist Henry Havelock Ellis published observations of some most peculiar behavior in hoofed domestic animals in his 1927 study that identified bulls, goats, sheep, camels and elephants as species known to practice masturbation. His observations, albeit second-hand, about the autoerotic habits of goats is a bit hard to swallow: 'I am informed by a gentleman who is a recognized authority on goats, that they sometimes take the penis into the mouth and produce actual orgasm, thus practicing auto-fellatio.'

So it seems from this brief survey of both recent and long-established findings on animal behavior that many of the sexual activities that might be have been labeled perverse by some – though risqué or even normal by others – can be found in one form or another in many wild animals. It is all just a part of what we now define as the wider spectrum of mammalian sexuality.

We have come a long way since the days of awkward bony claspers in primitive fishes, some 400 million years ago, to the highly pleasurable act of copulation we humans enjoy today. The last link in our chain of exploration, from the mechanism of sex to the evolution of sexuality, is to look at new discoveries in what goes on *after* the deed is done. Darwin thought that sexual selection was a major driver of evolutionary change, and he certainly wasn't wrong there. What he didn't know, though, is that natural selection often goes on inside the body of the inseminated female, a condition referred to by experts in reproductive physiology as 'sperm competition', and in some cases more aggressively as 'sperm wars'. This exciting field of study is whole new ball game in understanding evolution in action.

13

Sperm Wars: What Fossils Can't Tell Us

Nothing, or almost nothing, in animal reproduction
makes sense except in the light of sperm competition.

Professor Tim Birkhead, University of Sheffield

When Robin Baker's controversial book *Sperm Wars* hit the *New York Times*' bestseller list in 1997 it caused a media sensation. With its tales of human sperm killing opposition sperm from rival males, and kamikaze sperm using enzymes to wipe out their opposition even if it means sacrificing themselves, it created an image of miniature soldiers at war inside the uterus. However a review of the book's sensational claims by leading reproductive biology experts Tim Birkhead, Harry

Moore and Mike Bedford points out serious flaws in many of the key experiments used to demonstrate the assertions in the book. Much of them, they claim, is based on circumstantial evidence, and experiments could not always replicate the same results when repeated. Birkhead and his colleagues' review claimed that the public were 'misled on a grand scale'.

The take-home message from all this is that when an exciting new field of science emerges, such as the discovery of sperm competition in animals, care must be taken with designing experiments to get meaningful results rather than results that can be interpreted in spectacular but hard to prove ways. Despite this dampening down, sperm competition has become a hot new field in evolutionary biology since its modern debut by pioneers such as Dr Geoff Parker in the 1970s.

We have seen throughout this book how fossils can sometimes tell us about the shape and structure of reproductive organs in extinct groups of organisms, but the true modern focus of reproductive biology research is in fact happening at a cellular level. Unfortunately, this cellular information is not something likely to be expanded upon through future fossil finds due to the extreme improbability of such fine detail ever being preserved in even exceptional fossil remains. So instead, I'll attempt to provide a brief overview of how much of evolution, driven through reproductive adaptations, might have been shaped by the internal anatomy and physiological processes of our ancestors.

Charles Darwin may have recognized the value of sexual selection in his work on evolution, but he wrongly saw females as mostly passive players. It's true that sperm competition is prevalent, with sexual selection influencing the survival of the species well after copulation in many species ranging from snails and giant squid through to mammals. There have been many insightful breakthroughs in our understanding of the post-copulatory struggles that go on inside organisms as sperm from different male sources often compete to fertilize the female's eggs. But it is not just a case of which sperm is fastest or strongest, but in some cases which can last the longest in storage, or what other anatomical features have evolved inside the female's reproductive tract to prevent the sperm of some males from getting to the eggs.

Back in 1979, a study of the damselfly by Jonathan Waage of Brown University was the first time a biologist demonstrated the existence of a creature that possessed a penis-like structure adapted for both sperm transfer and sperm removal, with a series of long stiff hairs that could scrape out the sperm from the previous mating. Ghost crabs and spider crabs also demonstrate this approach but do it in a very cunning way. An initial mating transfers seminal fluid to the female, but no sperm, as the fluid acts like a hardening resin which first pushes any sperm from a previous male to the back of the sperm store and then seals it solid with a plug. Then, at the next mating, the male furnishes her with fresh sperm that is now at the top of the sperm reservoir, ready to fertilize her eggs.

Females of the humble fruit fly (*Drosophila melanogaster*) have long been known to mate with multiple partners and utilize only the sperm from the last male, called second male sperm precedence. In 1999 Catherine Price and her colleagues from the University of Chicago were able to crossbreed flies to express a particular type of green fluorescence on the tails of sperm, and also identified paternity from certain eye colors. Using these traits they were able to identify the sperm from certain males and then label and count sperm after matings. While this was no doubt detailed work, it was made slightly easier because these flies have enormously large sperm. Indeed, some fruit flies (*Drosophila bifurcata*) have single sperm that can be up to 2.3 inches (58 millimeters) long in an adult male fly only one-twentieth of an inch (1.5 millimeters) long. In these species the tails of the sperm are tightly wound around for delivery to the female, but nonetheless they are so large they are still visible to the naked eye as tiny white flecks.

The experiments by Price and her team demonstrated that the male sperm has the ability to displace and incapacitate previous reservoirs of sperm from earlier matings. The female flies have three receptacles inside to store sperm and in which all the sperm from the various male partners mixes together and is not compartmentalized, yet the studies demonstrated that the second male sperm inhibited the use of the previous male sperm without displacing it (technically called incapacitation). Sperm counts before and after fertilization of eggs and genetic matching of paternal lines gave the data needed to show how this second male precedence worked. The fresh sperm has an in-built ability to displace or incapacitate older sperm inside the storage vesicles. Such research is valuable to demonstrate the actual mechanism of sperm displacement and competition at work in living organisms, and how our interpretation of genetic traits in such organisms are much at the mercy of mating behaviors and post-mating sperm competition.

For many years it was thought that longer, faster sperm should have an edge in the evolutionary stakes over slower sperm, especially in external environments such as fish spawning sites in open water. A study of 29 species of Lake Tanganyika cichlid fishes, which are part of a rapid evolutionary diversification event (many new species having evolved from the ancestral stock in a relatively short time), were studied by researchers at the University of Western Australia to test this idea. They found a positive correlation between sperm length and swimming speed, as well as a correlation between increased female mating behavior and selecting for faster, more mobile sperm.

People have known for more than a thousand years that chickens could store sperm and release it months after their mating with roosters, but no-one knew how they were doing it until 1875. The discovery that chickens had little pouches near the uterus–womb junction to store sperm and release it as they needed to fertilize their eggs was a big breakthrough for Danish scientist Peter Tauber in that year. He had spent 25 years working on his doctorate on chicken fertilization, but unfortunately a serious disagreement with his supervisor meant his thesis was never passed. He is often not credited with the discovery, which is instead attributed to South African biologist C. van Drimmelen, who published his own work in 1946.

It's not surprising that many of the key studies about sperm competition have been done on birds. Birds are easily observed and have relatively quick gestation periods so that their young can also be studied and genetically tested. Birds use a variety of techniques to make sure theirs is the dominant sperm. Many birds are either polyandrous (females have multiple male partners) or a combination of monogamous and polyandrous, depending on the season. In species where males guard a fertile partner, copulations are infrequent, but when males are not around to look after their mates, copulation frequency rises.

The dunnock (*Prunella modularis*) is a small North American bird, and its often polyandrous females will copulate with two male partners over 250 times for each clutch of eggs. Other monogamous pairs of dunnocks will only mate about 50 times for each egg clutch. Pre-mating behavior in either case can involve males pecking at the swollen red cloaca of the female to encourage her to emit a drop of semen, most likely from a previous mating, before the male mounts her. What follows next is one of the fastest copulations on record: the male dunnock can offload his sperm in just a tenth of a second of intercourse.

Initially scientists assumed that males controlled the mating behavior in birds, but now it seems the females are the ones deciding when copulations should take place; in doing so they are improving the quality of sperm that fertilizes their eggs. Recently, an exciting discovery was made by a team from Macquarie University in Sydney while studying the beautifully colored Australian Gouldian finches (*Erythrura gouldiae*): it seems that the female finches can specifically target males that are compatible with their own genes. The normally socially monogamous females, when exposed to extra males in experimental cage conditions, took advantage of the additional opportunity to mate and take in sperm only from males compatible with their best chance for viable offspring. Then tests analyzed the chicks' DNA and showed that the females were releasing sperm from the second male to give them a disproportionately high number of fertilized eggs, even after their monogamous mate had been returned to further copulate with them.

Many examples of the complex behaviors involved in bird sperm competition are given by Tim Birkhead in his book edited with Anders Pape Møller *Sexual Selection and Sperm Competition* (1998).

Sperm competition in mammals is perhaps best exemplified by the ground squirrels of North America (aptly named *Spermophilus*, meaning 'sperm lovers'). These fluffy little mammals with cute tails sport a horrendous-looking penis armed with knife-edge projections, similar to those seen on the agouti, the South American rodent described previously. The penis is designed to cut through and remove copulatory plugs from previous insemination events. The murids (rats and mice) also use this method of sperm competition, with some able to take it a step further and create double plugs. In the desert kangaroo rat (*Dipodomys desertus*) of the Mojave and Sonoran Deserts in southwestern USA, females can develop a second plug by mixing accumulated vaginal secretions together with sloughed epithelial cells; when combined with the male's copulatory plug after copulation, the female is able to better control the quality of sperm that fertilizes her eggs. Meanwhile the North American meadow vole (*Microtus pennsylvanicus*) has been found to produce more sperm in each copulation if he smells another male in the neighborhood. And there are many more cases of mammals using ingenious methods of sperm competition.

So what of sperm competition in us primates? We have discussed earlier how gorillas, humans and chimpanzees have different-sized genitalia and produce different amounts of sperm relative to their mating habits. Chimpanzees have the largest testicles of all primates because they live in promiscuous societies, so the males need more sperm at each mating to try to out-compete the last male who copulated with the female. Gorillas and humans, on the other hand, have smaller testes and need smaller ejaculates, as they generally mate with only one female in each estral cycle. A study of the sperm morphology of primates has even shown that the midpiece part of sperm flagellum (or tail) in primates with multiple-partner mating systems (such as chimpanzees) are much thicker, enabling greater loading and sperm mobility. This is evidence that even sperm evolves and changes its physical form as a result of sexual behavior.

There may have been controversy surrounding Robin Baker's 1997 book on sperm wars, but he and Mark Bellis published some innovative research ideas as well.

In a classic 1993 paper they suggested that the female orgasm evolved as a method to 'suck up' more seminal fluid into the cervix for increased chance of fertilization. They observed that during times of infidelity, females changed the pattern of their orgasm to favor sperm from the 'extra-pair' male (that is not their primary partner), presumably raising his chance of fertility success over the long-term or primary partner's. This was highly controversial stuff.

In order to test Baker and Bellis's ideas and other related behavior, Dr Gordon Gallup and his colleagues from the State University of New York used a sample of 652 college students, their 2006 paper reported that the incidence of 'double mating' by females (admitted by one in four females in the test group) was enough to statistically make semen displacement a useful adaptation for males. They also set out to test the hypothesis that the human penis might have evolved to enable males to substitute their semen for that of their rivals. That research involved different-shaped artificial penises being inserted in a range of latex vaginas to try to test scientifically if the human male penis had evolved with its unusually well-developed glans in order to displace semen from a previous mating. Their results showed that it does indeed achieve this purpose to some extent in certain shape and size configurations.

Another study on human sperm competition involved a group of 305 males from a university and surrounding community. The aim was to show that human males at greater risk of mate infidelity used a higher degree of mate-retention tactics, such as concealing their partner by not introducing them to rival males, or trying to secure the bond with gifts such as jewelry and other things. In such experimental trials, the likelihood of sperm competition was determined by a series of questions that enabled researchers to see how each male rated his female partner in terms of physical attractiveness and personality traits. The test held true; in many cases males in weaker relationships were shown to be trying harder to keep their partners away from rival males.

Finally, a study by a team at the University of Western Australia tried to critically examine the claim that sperm competition is an important selection pressure in humans. They recruited 222 men and 194 women to complete a survey on their sexual behavior. From this they found that 28 per cent of the men and 22 per cent of the women had extra-pair copulations. Their study found ultimately that there was no substance in previously reported claims that men who engaged in extra-pair copulations had larger testes than men in monogamous relationships. Most significantly, their conclusion was that as the risk of a regular couple having a child to another extraneous male is only about 2 per cent for humans in our modern western societies, this creates a relatively low risk of sperm competition compared with our primate relatives. So although emerging studies claim that there are mechanisms in operation for human sperm competition, the reality is that it is more behavioral (for example, how we act towards our partners) than directly physiological (for example, the shape of the penis) in our species.

From all this recent research it's clear that much has transpired in our evolution from the humble beginnings of male organs in ancient fishes through to sperm manipulation via mating behavior or direct physiological conduits inside the female. But are there direct links from those ancient fishes to us? Can we still find the relict structures embodied and buried deep within our own human forms which first originated in these primordial armored fishes?

14

From Clasper to Penis: We've Come a Long Way, Baby

*If I were to give an award for the single best idea anyone
has ever had, I'd give it to Darwin, ahead of Newton
and Einstein and everyone else. In a single stroke, the
idea of evolution by natural selection unifies the realm
of life, meaning and purpose with the realm of space,
time, cause and effect, mechanism and physical law.*

Daniel Dennett
Darwin's Dangerous Idea: Evolution and the Meaning of Life

The final paragraph of Peter Watson's *A Terrible Beauty*
– covering all the great ideas of the 20th century – hailed
evolution as *the* most important idea, one that changed

the way we humans think about ourselves. Evolution reframed humans as just another species derived from a vast, unbroken chain of DNA, stretching from the first reproducing microbes almost four billion years ago to the great diversity and beauty of today's living creatures. Our human body form has evolved over the past 500 million years since the first fishes appeared through the slow acquisition of various anatomical novelties, the sequence of which is forever captured, frozen in time, by fossils. This is a true and humbling revelation about one's self. And there is a legacy inside all of us that demonstrates our direct links to the first armored fishes, and to their extraordinary methods of mating.

Today when we want to discover something new about our evolutionary history we can do two things. Firstly, we can go into the field and find fossils that reveal new data showing how the primitive pattern of bones may have transitioned from one form to another. Fossils like *Tiktaalik* or *Materpiscis* are good examples of specimens that revealed significant new information about our distant evolution. The second method is to examine the distant history that is entrapped in our own genes within our chromosomes. Our DNA, which is in effect tiny pieces of a gene, is a treasure trove of past history about our evolutionary development. This field of science is termed 'evolutionary developmental biology', or simply 'evo-devo' by us devotees. Evo-devo involves studying the development of animals through analyzing their embryonic stages and trying to determine which genes play which roles in the process.

The really big breakthrough of modern-day evolutionary biology has been the discovery of Homeobox genes (often shortened to Hox), which determine the sequence or organization of how body parts are built. In essence they are a combined toolkit and blueprint for the way our bodies develop, and once the egg has been fertilized and our cells start dividing, they go into action.

Big breakthroughs in science of truly immense significance to humanity are often recognized with Nobel prizes, and in 1995 the prize for medicine was awarded to professors Edward B. Lewis, Christiane Nüsslein-Volhard and Eric F. Wieschaus 'for their discoveries concerning the genetic control of early embryonic development', or Hox genes. Their search for the elusive organizer genes that scientists had thought played some role in setting up sequences in development of embryos started in the 1970s. Dosing fruit flies with mutation-inducing chemicals to see which genes had what effect in the developing fly embryo, the team had identified by about 1980 the key genes that determined the fly's body plan: they called them Homeobox genes, or Hox genes. Ed Lewis from Caltech determined that Hox genes were arranged along the chromosome in the same order as the body parts they control. The mind-boggling part of it all was when they discovered that many of the genes found in the fruit fly existed in all other animals, and played a similar role in the development of the same parts of the body, whether it be in sea-urchins, frogs, mice or humans. This find beautifully exemplified the common evolutionary ancestry of all living things.

So how does all this show that placoderms are directly related to us? Well, back in 1993, Cliff Tabin at Harvard University discovered that a certain Hox gene he called 'hedgehog' was involved in the development of our limbs. When he later discovered the specific hedgehog gene for a chicken, being a slightly different variant, he named it 'sonic hedgehog', after the computer game. Sonic hedgehog, abbreviated to 'shh', plays a vital role in the development of the limbs of vertebrates, in particular the fingers on our hands. After they had identified shh in all other animals with fingers, from frogs to humans, another research group lead by Professor Neil Shubin

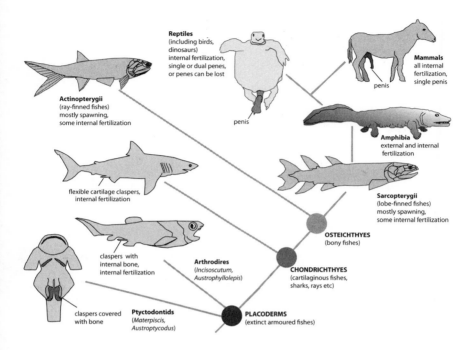

This diagram summarizes the evolution of vertebrate male reproductive systems, from paired bone-covered claspers (placoderms) to the single penis found in mammals, some birds and reptiles. Throughout our evolution the penis has sometimes been lost and then re-emerged many times in different lineages, according to need. The same Hox genes operating to form the claspers in sharks and ancient placoderms appear to have controlled this patterning right through to us humans. (John Long)

was able to trace the same gene in fishes (which clearly don't have fingers). Dr Randy Dahn and Neil Shubin also identified it in sharks as well as bony fishes. In essence they found that just because shh is involved

From Clasper to Penis: We've Come a Long Way, Baby

in digit development didn't mean it wasn't useful in developing the limb as a whole.

In 2007 Dahn's team published an exciting paper in *Nature* announcing that, after transplanting a bead containing shh from a mouse into the fin of a developing skate, they discovered that the gene played the exact same role in pattern development as it did in the mouse: a mutant fin skeleton was formed in the skate that developed new fin rods of a different shape to create a complex, almost hand-like pattern in the ray's pectoral fin.

We now know from these and other evolutionary developmental biology experiments that all limbs – indeed all appendages, whether limbs or fins – need shh to develop them. As shh is found in the most primitive of all living jawed fishes, sharks and rays, it signals back to the beginning of our own evolution when limbs were in the making, their blueprint deep inside the DNA of ancient fish but not yet able to release its full potential.

The combined study of evo-devo with the study of fossils has revolutionized our understanding of evolutionary transitions from fishes to the first land animals. The almost symmetrical digit patterns in the hands and feet of early tetrapods (as opposed to the pattern in later tetrapods which have five digits all difference size) are better explained in the light of Hox genes. Thus the timing of when certain genes first became activated to develop certain patterns in skeletons can be matched with actual fossil data.

And evo-devo is also a growth area that demonstrates how knowledge of something as obscure as Devonian fish anatomy can also play a major role in developing research programs in regenerative medicine that will one day greatly benefit us all. Dr Catherine Boisvert, who studied the fins of the Devonian fish *Panderichthys* for her doctorate work at Uppsala University, is now a leading researcher investigating shark embryonic development and gene activation at Monash University's new Australian Regenerative Medicine Institute. The study of embryonic development in these fishes by Boisvert which shows when the genes first became activated in an evolutionary sense may help future researchers develop new limbs, muscles or even spinal tissue directly. Such studies hold out great hope for future treatments by genetic regeneration of lost tissues or diseased organs.

Other genes that play an important role in limb development are the Hoxd series, in particular Hoxd9 to Hoxd13. A 2007 landmark paper by Renata Freitas and colleagues from the University of Florida studying shark limbs announced that early limb development started with a sequential activation of Hoxd13, which set in motion the mechanism of building the limb from near the shoulder girdle down to the mid section of the fin, or forearm equivalent in our arms. Then unexpectedly a second phase of activity kicks in (called a biphasic activation) whereby Hoxd12 and Hoxd13 are re-expressed along the distal or outer edges of the fin in the shark. They concluded that this sequence of activation, just like finding a new fossil, gives us a new evolutionary clue, a new pattern for early limb development in sharks, bony fishes and all later animals. Reading this particular paper, my eyes lit up when I saw the beautiful color photos of tiny shark embryos with their claspers illuminated by dyes indicating gene activation. It seems Hoxd12 is active in clasper formation, and Hoxd13 is particularly important as it activates the end of the clasper formation whilst also activating the cloacal urogenital area.

So how does this relate back to us humans? A study of developmental biology in mice by Martin Cohn published in 2004 showed that the genital tubercles (the bud from which genital organs will develop – see image in picture section) and limbs are both activated simultaneously by Hoxd13. He concludes that:

'Limbs and external genitalia undergo many similar morphogenetic processes, and ... the same molecular mechanisms may operate during development of the limb bud and the genital tubercle.'

This was a revelation for me. Here was the first piece of modern developmental biology data tying the origins of my placoderm claspers directly to the mammalian penis. Whereas the actual substances of how genitals are made might differ throughout the evolution of vertebrates, the genes that develop them appear to be the same and can be traced right back to the very origins of limbs. Like building a house out of bricks, wood or straw, as long as the blueprint is the same, the same kind of structures will be built, even if the materials used are different. Both placoderm and shark claspers are formed from a part of the developing pelvic girdle which was predestined to one day become the legs of us humans.

So initially, the reproductive structures of early jawed fishes were paired as part of the leg or hind limb pattern of bones. After paired fins had transformed into legs, the claspers were lost, but instead other paired reproductive structures merged from the urogenital plate, such as the hemipenes of lizards and snakes. In subsequent evolutionary radiations, the paired units became unnecessary as only one mating organ was required to do the job properly, so the single penis emerged, so to speak, as the dominant male reproductive organ in higher vertebrates.

Of course, losing the penis altogether happened several times in vertebrate evolution, as in some primitive reptiles (the tuatara) and the vast majority of flying birds (passerines), when for one reason or another an alternative reproductive strategy evolved. Just as limbs can be lost several times independently throughout vertebrate evolution (legless lizards, caecilians, snakes have all lost legs independently) or even several times within the many species of a single genus (as in Australian *Lerista* lizard species), so penis loss was apparently no big deal for some lineages of vertebrates.

We humans sometimes speak about 'getting a leg over' when referring to having sex, but for placoderms and it was more about 'getting a leg in'. Our humble mammalian penis might not look like it has a long evolutionary history, but indeed it has a deep developmental pedigree that can now be traced back in time to the very origins of arms and legs in all backboned animals.

So when you are next making love to that special person, and enjoying all the physiological pleasures afforded by our anatomy, give a little cry of joy for the ancient armored placoderm and all that it has given us. Because of some strange twist of biological fate, we have kept one of the most interesting parts of our reproductive anatomy from our archaic evolutionary history when other lines of animals managed to do perfectly well without it.

EPILOGUE

The Greatest Mystery of Biology

My story about the origins of vertebrate sex and what fossils reveal is now concluded, but the research still goes on. Each expedition to the site scours the dusty paddocks up at Gogo and other sites to try to find more spectacular fossils, hoping that one day we will uncover something truly amazing that either confirms an earlier theory or gives us reason to doubt earlier conclusions, perhaps even leading us on a totally new line of investigative thought. That's just how science works: it's not about who's right or wrong but what the truth is, and how we can best determine the value of new discoveries to understanding the bigger picture of our evolution.

Whilst researching the background material for this book I became distracted, as usual, and at one point followed a trail of the historical background to our understanding of sex. My reading uncovered amazing tales, such as the first discovery of sperm in 1677 by Antonie Van Leeuwenhoek, who incidentally built the first microscope powerful enough to see sperm. I also found out that a famous Danish scientist named Nicolas Steno (1638–1686), who studied fossils and founded the law of superposition about younger rocks lying on top of older rocks, also discovered the ovary whilst dissecting a shark. The more I searched, the more fascinating information was uncovered – but that's a whole other book.

The key to understanding how sex works is to understand fertilization. Today we know well that a sperm must fertilize an egg to start the biological process of making a new baby, and since the days of ancient Greece people have known that copulating will sometimes bring forth a new child, but how this happened was a total mystery. Some, like the ancient Greeks, believed that the 'female seminal fluids' formed the foetus whereas the males' seminal fluids only provided nourishment, and this idea prevailed for over a thousand years.

Sir Thomas Aquinas, a prolific writer and philosopher of the thirteenth century, thought that the active power was the froth in the semen, which had a special heat of its own derived not from the soul of man but from the action of the heavenly bodies. Indeed throughout the Middle Ages, intellectual battles waged between the two sides of biological thought on the matter, the spermists and the ovists: some thought that entire human beings were preformed and bundled up inside each male sperm, whereas others thought the egg was the sacrosanct body containing the undeveloped embryo and that the male only played a role in supplying fluid to nourish it.

Strange experiments were performed in the late 1700s to try to determine the exact roles that male and female seminal fluids played. The famous Italian abbot Lazzaro Spallanzani made little pairs of taffeta pants and fixed them onto frogs before watching them mate. Unsurprisingly to us, he found that frogs wearing his special pants could not fertilize the female's eggs. He then took the semen from the frogs' pants, applied them to the eggs and discovered they would develop into tadpoles. Spallanzani was also the first scientist to artificially inseminate a dog, so his work proved that male semen was a necessary part of fertilization. But after a lifetime of precise experimental work in which he made many groundbreaking discoveries, he concluded that the little animalcules (as he called them – we call them 'sperm') had nothing at all to do with fertilization.

If asked what the greatest mystery of biology was, many people assume the answer to be evolution. Yet Charles Darwin published a book espousing his theory of evolution in 1859, some 17 years before the two-millennia-old enigma of what causes fertilization in animals and humans was finally solved.

Not long ago I gave my standard lecture about the origins of sex based on our fossil discoveries to a series of learned audiences at museums, colleges and universities throughout the USA. I posed one question at the end of each lecture: who was the great scientist who first discovered the secret of fertilization? Not one person in any of these well-educated crowds gave me the right answer. It seemed to me that there is even more of this fascinating story of sex to be told, from the ancient Greeks to modern researchers developing such things as in-vitro fertilization. It seems my unexpected journey into the world of sex is not over yet.

And by the way, the name of the great genius who is credited with the discovery of how fertilization really works is Oscar Hertwig (1849–1922), who was Professor of Biology at Berlin University.

NOTES

Preface

Brain in orgasm: Vaitl, D., Birbaumer, N., Gruzelier, J., Jamieson, G.A., Kotchoubey, B., Küber, A., Lehmann, D., Miltner, W.H.R., Ott, U., Pütz, P., Sammer, G., Strauch, I., Strehl, U., Wackerman, J., and Weiss, T. (2005) 'Psychobiology of altered states of consciousness', *Psychological Bulletin*, 131: 98–127.

Sada Abe: Wikipedia http://en.wikipedia.org/wiki/Sada_Abe.

Charles Darwin: Darwin, C. (1871) *The descent of man*, various editions.

Biodiversity estimates: en.wikipedia.org/wiki/Biodiversity#cite_note-54.

Tetrapod evolution from fishes: Long, J.A. (2010) *The Rise of Fishes: 500 million years of evolution*, Johns Hopkins University Press, Baltimore, UNSW Press, Sydney.

Human bottleneck one million years ago: Huff, C.D., Xing, J., Rogers, A.R., Witherspoon, D. & Jorde, L.B. (2010) 'Mobile elements reveal small population size in the ancient ancestors of *Homo sapiens*', *Proceedings of the National Academy of Sciences* 107: 2147–52.

DNA: For a simple explanation of DNA see http://en.wikipedia.org/wiki/DNA

Fossils: For an explanation of fossils see Long, J.A. (2011) *The Rise of Fishes: 500 million years of evolution*, Johns Hopkins University Press, Baltimore, UNSW Press, Sydney.

Jared Diamond: Diamond, J. (1997) *Why is Sex Fun?: The evolution of human sexuality*, Basic Books, New York.

Susan Windybank: Windybank, S. (1991) *Wild Sex: Way beyond the birds and the bees*, Reed Books, Australia, St Martin's Press, New York.

1 The Machismo of the Argentine Duck

Argentine ducks: McCracken, K.G. et al (2001) 'Are ducks impressed by drake's display?', *Nature* 413: 128.

Blue whale penis: www.abs-cbnnews.com/lifestyle/11/18/09/how-long-blue-whales-penis-0

Dr Dickinson: Dickinson, R.L. (1940) *The Sex Life of the Unmarried Adult*, Vanguard Press, New York; http://robertdickinson.blogspot.com/.

Professor Short: Short, R. (1979) 'Sexual selection and its component parts, somatic and genital selection as illustrated by man and the great apes', *Advances in the Study of Behavior* 9: 131–58.

Explosive duck penis: Brennan, P., Clark, C. and Prum, R., 'Explosive eversion and functional morphology of the duck penis supports sexual conflict in waterfowl', *Proceedings of the Royal Society of London B*, 277 (1686): 1309–14.

For background about Gogo fishes and my Gogo expeditions: Long, J.A. (2006) *Swimming in Stone: The amazing Gogo fossils of the Kimberley*, Fremantle Arts Centre Press, Perth.

For images of Gogo sites and spectacular fossils: Long, J.A. (1988) 'Late Devonian fishes of the Gogo Formation, Western Australia', *National Geographic Research and Exploration* 4: 436–50.

Long, J.A. (2011) *The Rise of Fishes: 500 million years of evolution*, Johns Hopkins University Press, Baltimore, UNSW Press, Sydney.

Ctenurella gardineri: Miles, R.S. & Young, G.C. (1977) 'Placoderm interrelationships reconsidered in the light of new ptyctodontids from Gogo, Western Australia', in Andrews, S.M., Miles, R.S. and Walker, A.D. (eds), *Problems in Vertebrate Evolution*, Linnean Symposium Series 4: 123–98

Placoderm morphology: Long, J.A. (2010) *The Rise of Fishes: 500 million years of evolution*, Johns Hopkins University Press, Baltimore, UNSW Press, Sydney.

Tor Ørvig's description of *Ctenurella*: Toombs, H.A. (1948) 'The use of acetic acid in the development of vertebrate fossils', *Museums Journal* 48: 54–5.

New genus *Austroptyctodus*: Long, J.A. (1997) 'Ptyctodontid fishes from the Late Devonian Gogo Formation, Western Australia, with a revision of the European genus *Ctenurella* Ørvig 1960', *Geodiversitas* (Paris Museum of Natural History) 19: 515–56. See also Long (2011) for information about placoderms, palaeoniscids and other Devonian fishes.

2 The Mother of All Fossils

Acid preparation: Toombs, H.A. (1948) 'The use of acetic acid in the development of Vertebrate fossils', *Museums Journal*, 48: 54–5.

Curt Teichert and the discovery of Gogo: Long, J.A. (2006) *Swimming in Stone: The amazing Gogo fossils of the Kimberley*, Fremantle Arts Centre Press, Perth.

Hypotheses about where placoderms fit in phylogenetically: Young, G.C. (2010) 'Placoderms (armored fish): Dominant vertebrates of the Devonian Period', *Annual Reviews in Earth and Planetary Sciences* 38: 523–50; Brazeau, M. (2009) 'The braincase and jaws of a Devonian "acanthodian" and modern gnathostome origins', *Nature* 457: 305–8; Goujet, D. and Young, G.C. (2004) 'Placoderm interrelationships', in *Recent Advances in the Origin and Early Radiation of Vertebrates* (eds Arratia, G., Wilson, M. and Cloutier, R.), Dr Freiderich Pfeil, Munich, 2004: 109–26.

Delphydontos: Lund, R. (1980) 'Viviparity and intrauterine feeding in a new holocephalan fish from the Lower Carboniferous of Montana', *Science*, 209: 697–9.

Male and female ptyctodontid: Miles, R.S. (1967) 'Observations on the ptyctodont fish, *Rhamphodopsis* Watson', *Zoological Journal of the Linnean Society* 47: 99–120.

3 The Ptyctodontid Kind of Congress

Great white sharks: Pratt, H.L. Jr, in *Great White Sharks: the biology of* Carcharodon carcharias (eds Klimley, A.P. and Ainley, D.G.), Academic Press Inc.

Mating in sharks and stingrays: Chapman, D.D., Corcoran, M.J., Harvey, G.M., Malan, S., and Shivji, M.S., (2003) 'Mating behavior of southern stingrays, *Dasyatis americana (Dasyatidae)*',

Environmental Biology of Fishes 68: 241–5; Chapman, D.D., Prodohl, P.A., Gelsleichter, J., Manire, C.A., and Shivji, M.S., (2004) 'Predominance of genetic monogamy by females in a hammerhead shark, *Sphyrna tiburo*: Implications for shark conservation', *Molecular Ecology* 13: 1965–74; Tricas, T.C. and Le Feuve, E.M. (1985) 'Mating in the reef white-tipped shark *Triaenodon obsesus*, *Marine Biology*, 84, 233–7.

Professor Watson: Watson, D.M.S. (1934) 'The interpretation of arthrodires', *Proceedings of the Zoological Society of London*, 3: 437–64; Watson, D.M.S. (1938) 'On *Rhamphodopsis*, a ptyctodontid from the Middle Old Red Sandstone of Scotland,' *Transactions of the Royal Society of Edinburgh* 59: 397–410'.

Dr Miles: Miles, R.S. (1967) 'Observations on the ptyctodont fish, *Rhamphodopsis* Watson', *Zoological Journal of the Linnean Society* 47: 99–120.

Miles and Young: Miles, R.S. and Young, G.C. (1977) 'Placoderm interrelationships reconsidered in the light of new ptyctodontids from Gogo, Western Australia', in Andrews, S.M., Miles, R.S. and Walker, A.D. (eds), *Problems in Vertebrate Evolution*, Linnean Symposium Series 4: 123–98.

4 Announcing Fossil Sex to the Queen

Mooney and Kirshenbaum: Mooney, C. and Kirshenbaum, S. (2009) *Unscientific America: How scientific illiteracy threatens our future*, Basic Books, New York.

Images of the Royal Institution re-opening: www.mayfair.org.uk/ blog/tag/royal-institution; www.talktalk.co.uk/news/daily/ photos/galleries/view/daily/20080528/browse/6.

Announcing the news to the world – media coverage: news.nationalgeographic.com/news/2008/05/080528- mother-fossil.html; www.telegraph.co.uk/science/science- news/3343083/Fish-fossil-is-oldest-to-have-fun-sex.html;

Australasian Science: Long, J.A. (2008) 'Mother fossil', *Australasian Science* 29 (6): 16–18.

Discover top 100: discovermagazine.com/2009/jan/092

World's oldest live birth: *The Guinness Book of World Records 2010*, Guinness World Records, London, p. 54.

Scientific American cover story: Long, J.A. (2011) 'Dawn of the deed', *Scientific American* January 2011: 34–9.

5 Paleozoic Paternity Problems

Incisoscutum: Dennis-Bryan, K. and Miles, R. (1981) 'A pachyosteomorph arthrodire from Gogo, Western Australia,' *Zoological Journal of the Linnean Society* 73: 2 13–58.

Ctenurella: Ørvig, T. (1960) 'New finds of acanthodians, arthrodires, crossopterygians, ganoids and dipnoans in the upper Middle Devonian Calcareous Flags (Oberer Plattenkalk) of the Bergisch-Paffrath Trough (Part 1)', *Palaontologische Zeitschrift* 34, pp. 295–335.

South African placoderms: Long, J.A., Anderson, E., Gess, R. and Hiller, N. (1997) 'New placoderm fishes from the Upper Devonian of South Africa', *Journal of Vertebrate Palaeontology* 17: 253–68.

New phyllolepids for conference proceedings: Long, J.A. (1984) 'New phyllolepids from Victoria and the relationships of the group', *Proceedings of the Linnean Society of New South Wales*, 107: 263–308.

Second *Nature* paper: Long J.A., Trinajstic, K. and Johanson Z. (2009) 'Devonian arthrodire embryos and the origin of internal ferilsation in vertebrates', *Nature* 457: 1124–7; Ahlberg, P.E. (1989) 'Fossil fishes from Gogo', *Nature* 337: 511–12.

6 Finding the Daddy Fish

W.H. Leigh-Sharpe: Leigh-Sharpe, W.H. (1920) 'The comparative morphology of the secondary sexual characters of elasmobranch fishes. Memoir I', *Journal of Morphology* 34: 245–65; Leigh-Sharpe, W.H. (1921) 'The comparative morphology of the secondary sexual characters of elasmobranchs fishes. Memoir II', *Journal of Morphology* 35: 359–80; Leigh-Sharpe, W.H. (1924) 'The comparative morphology of the secondary sexual characters of elasmobranchs fishes. Memoirs VI and VII', *Journal of Morphology* 39: 553–77; Leigh-Sharpe, W.H. (1926). 'The comparative morphology of the secondary sexual characters of elasmobranch fishes. The claspers, clasper siphons, and clasper glands together with a dissertation on the Cowpers glands of Homo, Memoir XI',

Notes

Journal of Morphology 26: 349–58. For information about Leigh-Sharpe's life and works see Damkaer, D.M. and Merrington, O.J. (2005) 'William Harold Leigh-Sharpe (1881-1950): Teacher and copepodologist', *Journal of Crustacean Biology* 25(3):521-528.

Ahlberg, P.E., Trinajstic, K., Johanson, Z. and Long, J.A. (2009) 'Pelvic claspers confirm chondrichthyan-like internal fertilization in arthrodires', *Nature* 460: 888–9.

Kim Dennis-Bryan: Dennis-Bryan, K. (1987) 'A new species of eastmanosteid arthrodire (Pisces: Placodermi) from Gogo, Western Australia', *Zoological Journal of the Linnean Society* 90: 1–64; Dennis-Bryan, K. and Miles, R. (1981) 'A pachyosteomorph arthrodire from Gogo, Western Australia', *Zoological Journal of the Linnean Society* 73: 2 13–58.

Carcharhinus mating: see Tricas, T.C. and Le Feuve, E.M. (1985) 'Mating in the reef white-tipped shark *Triaenodon obsesus*, *Marine Biology*, 84, 233–7; Chapman, D.D., Corcoran, M.J., Harvey, G.M., Malan, S., and Shivji, M.S., (2003) 'Mating behavior of southern stingrays, *Dasyatis americana (Dasyatidae)*', *Environmental Biology of Fishes* 68: 241–5; Chapman, D.D., Prodohl, P.A., Gelsleichter, J., Manire, C.A., and Shivji, M.S., (2004) 'Predominance of genetic monogamy by females in a hammerhead shark, *Sphyrna tiburo*: Implications for shark conservation', *Molecular Ecology* 13: 1965–74.

Brad Norman and whale sharks: www.sciencedaily.com/ releases/2010/08/100824184754.htm; Schmidt, J.V, Chen, C.C, Sheikh, S.I., Meekan, M.G., Norman, B.M. and Joung, S.J. (2010) 'Paternity analysis in a litter of whale shark embryos', *Endangered Species Research* 12: 117–24.

MacArthur and Wilson: MacArthur, R.H. and Wilson, E.O. (1967) *The theory of island biogeography*, Princeton University Press, New Jersey.

Dr Bob Carr: Carr, R. (2010) Abstract in SVP Conference, Pittsburgh, 10–13 October.

7 Down and Dirty in the Devonian

Yseult Bridges: Bridges, Y. (1980) *Child of the Tropics*, Collins and Harvill Press, London, 1980, p. 29.

Fish copulation: Reebs, S.G. 'The sex lives of fishes', *www. howfishbehave.ca*

Sperm-drinking catfish: Khoda, M., Tanimura, M., Kikue-Nakamura, M. and Yamagishi, S. (1995). 'Sperm drinking by female catfishes: a novel method of insemination', *Environmental Biology of Fishes* 42: 1-6

Blue-head wrasse mating: Warner, R.R., Robertson, D.R. and Leigh Jr., E.G. (1975) 'Sex change and sexual selection', *Science* 190, 633–8.

Video of *Incisoscutum*: http://www.youtube.com/watch?v=qkFZYIgz37Q

8 At the Dawn of Archaic Sex

Reg Sprigg: Weidenbach, K. (2008) *Rock Star: The story of Reg Sprigg – an outback legend*, East Street Publications, Hindmarsh, SA.

Ediacaran fossils: Glaessner, M. (1958) 'New fossils from the base of the Cambrian in South Australia', *Transactions of the Royal Society of South Australia*, 81: 185–89.

Charnwood fossils: Ford, T. (1958) 'Precambrian fossils from Charnwood Forest', *Proceedings of the Yorkshire Geological Society*, 31: 211–17.

Debate about Ediacaran organisms: Fedonkin, M.A., Gehling, J.G., Grey, K., Narbonne, G.M. and Vickers-Rich, P. (2007) *The Rise of Animals: The evolution and diversification of the Kingdom Animalia,* Johns Hopkins University Press, Baltimore.

Funisia: Droser, M.L. and Gehling, J.G. (2008) 'Synchronous aggregate growth in an abundant new Ediacaran Tubular organism', *Science* 319: 1660–2.

London Times on *Funisia*: Smith, L. (2008) 'Fossils shed light on the history of sex', *London Times*, 21 March, www.timesonline. co.uk/tol/news/science/article3593959.ece

Biomarkers and oldest Eukaryotes: Brocks, J.J., Logan, G.A., Buick, R. and Summons, R.E. (1999) 'Archaean molecular fossils and the early rise of eukaryotes,' *Science* 285: 1033–36.

Rasmussen and colleagues: Rasmussen, B., Fletcher, I.R., Brocks, J.J. and Kilburn, M.R. (2008) 'Reassessing the first appearance of eukaryotes and cyanobacteria', *Nature* 455: 1101–4.

Genomic mutation: Keightley, P.D. and Eyre-Walker, A. (2000) 'Deleterious mutations and the evolution of sex', *Science* 290: 331–3.

Sarah Otto: Otto, S.P. (2009) 'The evolutionary enigma of sex', *American Naturalist*, supplement 174: S1–S14.

9 Sex and the Single Ostracod

BBC news spider: [web address P134]

Barnacle penis shape: Neufeld, C.J. and Palmer, A.R. (2008) 'Precisely proportioned: Intertidal barnacles alter penis form to suit coastal wave action', *Proceedings of the Royal Society of London*, B 275: 1081–7.

Ostracod penis: Siveter, D.J., Sutton, M.D. and Briggs, D.E.G. (2003) An ostracod crustacean with soft parts from the Lower Silurian,' *Science* 302: 1749–51.

Caribbean ostracods: Morin, J. and Cohen , A. (2010) 'It's all about sex: Bioluminescent courtship displays, morphological variation and sexual selection in two new genera of Caribbean ostracodes', *Journal of Crustacean Biology* 30: 56–67.

Harvestman fossils: Dunlop, J.A., Anderson, L.I., Kerp, H. and Hass, H. (2003) 'Palaeontology: Preserved organs of Devonian harvestmen', *Nature* 425: 916.

Living harvestmen behaviors: Buzato, B.A. and Macjado, G. (2009) 'Amphisexual care in *Acutisoma proximum* (Arachnida, Opiliones), a neotropical harvestman', *Insectes Sociaux* 56; Mora, G. (1990) Paternal care in a neotropical harvestman, *Zygopachylus albomarginis* (Archanida, Opiliones: Gonyleptidae)', *Animal Behavior* 39: 582–93.

Praying mantis cannibalism and bedbug sex: Birkhead, T. (2000) *Promiscuity: An evolutionary history of sperm competition*. Harvard University Press; Windybank, S. (1991) *Wild Sex: Way beyond the birds and the bees*, Reed Books, Australia, St Martin's Press, New York.

Harpactea sadistica: Řezáč, M. (2009) 'The spider *Harpactea sadistica*: Co-evolution of traumatic insemination and complex female genital morphology in spiders', *Proceedings of the Royal Society B*, 276 (1668): 2697–701.

10 Sex on the Beach

Grunion spawning: *Thompson, W.F.* (1919) 'The *spawning of the grunion (Leuresthes tenuis)*', *Californian Fish and Game* 5: 1–27; Jordan, D.S. (1926) 'The habits of the grunion', *Science* 63: 454; Mercieca, A., and Miller', R.C. (1969) 'The spawning of the grunion', *Pacific Discovery* 22: 26–7; Jordan 1926 and Mercaia and www.youtube.com/watch?v=B9h1tR42QYA

Early tetrapods: Niedźwiedzki, G., Szrek, K., Narkiewicz, P., Narkiewicz, M. and Ahlberg, P.E. (2010). 'Tetrapod trackways from the early Middle Devonian period of Poland' *Nature* 463: 43–8.

Tiktaalik: Shubin, N. (2008) *Your inner fish: A journey into the 3.5 billion year history of the human body*, Penguin, London.

Professor Jenny Clack: Clack, J.A. and Finney, S.M. (2005) '*Pederpes finneyae*, an articulated tetrapod from the Tournaisian of Western Scotland', *Journal of Systematic Palaeontology* 2: 311–46.

Tailed frog: Stephenson, B. and Verrell, P. (2006) 'Courtship and mating of the tailed frog (*Ascaphus truei*)', *Journal of Zoology* 259: 15–22.

Wood frogs: www.flickr.com/photos/21670394@N07/4510803394/.

Atrazine and frog sexuality: Hayes, T.B., Khoury, V., Narayan, A., Nazir, M., Park, A., Brown, T., Adame, L., Chan, E., Buchholz, D., Stueve, T., and Gallipeau, S. (2010) 'Atrazine induces complete feminization and chemical castration in male African clawed frogs (*Xenopus laevis*)', *Proceedings of the National Academy of Sciences* 107: 4612–17.

Caecilian spicules: Wake, M. (1988) 'Cartilage in the cloaca: Phallodeal spicules in caecilians (Amphibia: Gymnophiona),' *Journal of Morphology* 237: 177–86.

11 Dinosaur Sex and Other Earthshaking Discoveries

Beverly Halstead quotes: Fritz, S. (1998) '*Tyrannosaurus* sex: A love tail', *Omni*, February: 64–9.

Robert Plot: Plot, R. (1667) *Natural History of Oxfordshire*, Leon Lichfield, London.

Online opinion about dinosaur sex: dinosaurs.about.com/od/dailylifeofadinosaur/a/dinomating.htm

Notes

Henry the tuatara: news.bbc.co.uk/2/hi/asia-pacific/7850975.stm; news.bbc.co.uk/1/hi/world/asia-pacific/7850975.stm.

Bobtail lizard behavior: Bull, M.C., Cooper, S.J.B., Baghurst, B.C. (1998) 'Social monogamy and extra-pair fertilization in an Australian lizard, *Tiliqua rugosa*', *J. Behavioral Ecology and Sociobiology* 44 (1): 63–72.

Garter snake sex: Shine, R., Phillips, B., Waye, H., LeMaster, M., Mason R.T. (2001) 'Benefits of female mimicry to snakes', *Nature*, 414: 267.

Aldabran tortoises: Grubb, P. (1971) 'The growth, ecology and population structure of giant tortoises on Aldabra', *Philosophical Transactions of the Royal Society of London*, Series B, Biological Sciences, vol. 260, no. 836, A Discussion on the Results of the Royal Society Expedition to Aldabra 1967–68: 327–72.

Tortoise and mammal penis similarities: Kelly, D.A. (2004) 'Turtle and mammal penis designs are anatomically convergent', *Proceedings of the Royal Society of London B* (Supplement) DOI 10.1098/rebl.2004.0161; Jones, Frederic Wood (1915) 'The Chelonian Type of Genitalia', *Journal of Anatomical Physiology*, July, 49 (Pt 4): 393–406.

Marine reptile fossil viviparity: Maxwell, E. and Caldwell, M. (2003) 'First record of live birth in Cretaceous ichthyosaurs: closing an 80 million year gap', *Proceedings of the Royal Society of London B* (Supplement) 270: S104–7; Cheng, Y.-N., Wu, X., and Ji, Q. (2004) 'Triassic marine reptiles gave birth to live young', *Nature* 432: 383–6.

Komodo dragon clones: Watts, P.C., Buley, K.R., Sandserson, S., Boardman, W., Ciofi, C. and Gibson, R. (2006) 'Parthogenesis in Komodo dragons', *Nature* 444: 1021–2.

Carmelo Amelfi article: Amelfi, C. (2005) 'Tyrannosaurus sex', *Cosmos* 4.

Oviraptorosaur eggs: Sato, T., Cheneg, Y, Wu, X. M, Zelenitsky, D.W. and Hsiao, Y. (2005), 'A pair of shelled eggs inside a female dinosaur', *Science* 308: 375.

Fused tail vertebrae: Rothschild, B.M. and Berman, D.S. (1991) 'Fusion of caudal vertebrae in Late Jurassic sauropods', *Journal of Vertebrate Paleontology* 11: 29-36.

Bird origins from dinosaurs: Chiappe, L.M. (2007). *Glorified dinosaurs – The origin and early evolution of birds*, Wiley & Sons, New Jersey, USA; Long, J.A. and Schouten, P. (2008) *Feathered Dinosaurs: The origin of birds*, CSIRO Publishing, Melbourne.

Ostrich sex: Bolwig, N. (1973) 'Agonistic and sexual behavior of the African ostrich (Struthio cemelus)', *The Condor* 75: 100–5; Sauer, E.G. (1972) 'Aberrant sexual behavior in the South African ostrich', *The Auk*, 89: 717–37

Homosexual chinstrap penguins: Driscoll. E.V. (2008) 'Bisexual species', *Scientific American Mind*, June/July: 68–74. For other examples of avian homosexual behavior, Poiani, A. (2010) *Animal Homosexuality: A biosocial perspective*, Cambridge University Press.

12 We Ain't Nothing But Mammals

'The Bad Touch' rating: www.aolradioblog.com/2010/09/11/100-worst-songs-ever-part-three-of-five.

Diamond, J. (1997) *Why is Sex Fun?: The evolution of human sexuality*, Basic Books, New York.

Neanderthal births: Ponce de Leon, M.S., Golovanova, L., Vladimir Doronichev, V., Romanova, G., Akazawa, T., Kondo, O., Ishida, H. & Zollikofer, C.P.E. (2008) *Proceedings of the National Academy of Sciences* 105 (37): 13764–8.

Loss of genes: McLean, Cory Y., Reno, Philip L., Pollen, Alex A., Bassan, Abraham I, Capellini, Terence D., Guenther, Catherine, Indjeian, Vahan B., Xinhong Lim, Menke, Douglas B., Schaar, Bruce T., Wenger, Aaron M., Bejerano, Gill, and Kingsley, David M. (2011) 'Human-specific loss of regulatory DNA and the evolution of human-specific traits', *Nature* 471:216–19.

Penis structure: Kelly, D.A. (2007) 'Penises as variable-volume hydrostatic skeletons', *Annals of the New York Academy of Sciences* 1101: 453–63.

Agouti penis: Mollineau, W., Adogwa, A., Jasper, N., Young, K. and Garcia, G. (2006) 'The gross anatomy of the male reproductive system of a Neotropical rodent the Agouti (*Dasyprota leporina*)', *Anatomia Histologia Embryologia* 35: 47–52;

Notes

Todd, R. (1852) *Cyclopaedia of anatomy and physiology*, Vol. IV, Longman, Brown, Green and Longmans, London.

On genital shape and evolution: Eberhard, W.G. (1985) *Sexual selection and animal genitalia*, Harvard University Press, Cambridge, USA.; Hoskin, D.J. and Stockley, P. (2004) 'Sexual selection and genital evolution', *Trends in Ecology and Evolution* 19: 87–93.

New Guinea rodents: Lidicker Jr, W.Z., (1968) 'A phylogeny of New Guinea rodent genera based on phallic morphology', *Journal of Mammalogy* 49: 609– 643.

Fossil bacula: Koenigswald, W. von (1979) 'Ein Lemurenrest aus dem eozänen Ölschiefer der Grube Messel bei Darmstadt', *Palaontologische Zeitschrift* 53: 63–76; Dixson, A.F. (1987) 'Baculum length and copulation behavior in primates', *American Journal of Primatology* 13: 51–60; Dixson, A.F. 1987.

The sex life of echidnas: Morrow, G., Anderson, N.A. and Nicol, S.C. (2010) 'Reproductive strategies of the short-beaked echidna: A review with new data from a long-term study on the Tasmanian subspecies (*Tachyglossus aculeatus setosus*)', *Australian Journal of Zoology* 54: 274–82.

Review of animal homosexuality: Driscoll. E.V. (2008) 'Bisexual species', *Scientific American Mind*, June/July: 68–74.

Koalas: Feigea, S., Nilsson, K., Clive, J.C., Phillips, C.J.C. and Johnston, S.D. (2007) 'Heterosexual and homosexual behavior and vocalisations in captive female koalas (*Phascolarctos cinereus*)', *Applied Animal Behavior Science* 103: 131–45.; and Karen Nilsson, Lone Pine Koala Sanctuary, Brisbane, personal communication

Chinese fruit bat: Tan, M., Jones, G., Zhu, G., Ye, J., Hong, T., Zhou, S., Zhang, S. and Zhang, L. (2009) 'Fellatio by fruit bats increases copulation time', PLOS One October, 4 (10) e7595: 1–5.

Dyak fruit bat: Francis, C.M., Edythe, A.L.P., Brunton, J.A. and Kunz, T.H. (1994) 'Lactation in male fruit bats', *Nature* 367: 691–92.

Spotted hyenas: Cuhna, G., Wang, Y., Place, N.J., Lui, W., Baskin, L. and Glickman, S.E. (2003) 'Urogential system of the spotted hyena (*Crocuta crocuta erxleben*): A functional histological study', *Journal of Morphology* 256: 205–218.

Diamond and human sexual evolution: Diamond, J. (1997) *Why is Sex Fun?: The evolution of human sexuality*, Basic Books, New York; Ryan, C. and Jetha, C. (2010) *Sex at Dawn: The prehistoric origins of modern sexuality*, HarperCollins, New York.

Bonobos: Palagi, E., Paoli, T. and Tarli, S.B. (2004) 'Reconciliation and consolation in captive bonobos (*Pan paniscus*)', *American Journal of Primatology* 62: 15–30.

Work by Nancy Thompson Handler: in De Waal, F. (1995) 'Bonobo sex and society', *Scientific American* 272 (3).

Stump-tailed monkeys: Chevalier-Skolnikoff, S. (1976) 'Homosexual behavior in a laboratory group of stumptail monkeys (*Macaca arctoides*): Forms, contexts, and possible social functions', *Archives of Sexual Behavior* 5: 1–17.

Animal homosexuality: Poiani, A. (2010) *Animal homosexuality: A biosocial perspective*, Cambridge University Press, Melbourne.

Henry Havelock Ellis: Ellis, H.H. (1927) 'Studies in the psychology of sex', see Project Gutenberg, www.gutenberg.org/files/13610/13610-h/13610-h.htm.

13 Sperm Wars: What Fossils Can't Tell Us

Robin Baker: Baker, R. (1997) *Sperm Wars: The science of sex*, Basic Books.

Review of Baker: Birkhead, T.R., Moore, H.D.M. and Bedford, J.M. (1997) 'Sex, science and sensationalism', *TREE* 12: 121–2.

Damselfly: Waage, J.K. (1992) 'Dual function of the damselfly penis: Sperm removal and transfer,' *Science* 203: 916–18.

Fruit fly: Price, C.S.C., Dyer, K.A. and Coynes, J.A. (1999) 'Sperm competition between Drosophila males involves both displacement and incapacitation', *Nature* 400: 449–51.

Lake Tanganyika cichlid fishes: Fitzpatrick, J.L., Montgomerie, R., Desjardins, J.K., Stiver, K.A. and Balshine, S. (2009), Female promiscuity promotes the evolution of faster sperm in cichlid fishes', *Proceedings of the National Academy of Sciences* 106: 1128–32.

Chicken fertilization: Birkhead, T. (2000) *Promiscuity: An evolutionary history of sperm competition*, Harvard University Press; van Drimmelen, C.G. (1946) '"Sperm nests" in the oviduct of the domestic hen', *Journal of the South African Veterinary Medical Association*, 17: 42–52.

Gouldian finches: Pryke, S.R., Rollins, L.A. and Griffiths, S.C. (2010) 'Females use multiple mating and genetically loaded sperm competition to target compatible genes', *Science* 329: 964–7.

Sperm competition: Birkhead, T.R. and Møller. A.P. (eds) (1998) *Sperm Competition and Sexual Selection*, Academic Press, London.

Squirrels, Desert rats etc: Hartung, T.G. and Dewsbury, D.A. (1978) 'A comparative analysis of the copulatory plugs in muroid rodents and their relationship to copulatory behavior', *Journal of Mammalogy* 59: 717–23; Randall, J.A. (1991) 'Mating strategies of a nocturnal desert rodent (*Dipodys spectabilis*)', *Behavioral Ecology and Sociobiology* 28: 215–20; Mollineau, W., Adogwa, A., Jasper, N., Young, K. and Garcia, G. (2006) 'The gross anatomy of the male reproductive system of a neotropical rodent, the agouti (*Dasyprota leporina*), *Anatomia Histologia Embryologia* 35: 47–52.

Meadow Voles: del Barco-Trillo, J. and Ferkin, M. H. (2004) 'Male mammals respond to a risk of sperm competition conveyed by odours of conspecific males', *Nature* 431: 446–9.

Primate midpiece sperm: Anderson, M.J. and Dixson, A.F. (2002) 'Motility and the midpiece in primates', *Nature* 416: 496.

Baker and Bellis: Baker, R.R and Bellis, M.A. (1993) 'Human sperm competition: Ejaculate manipulation by females and afunction for the female orgasm', *Animal Behavior* 46: 887–909.

Gallup research: Gallup Jr, G.C., Burch, R. and Berens Mitchell, T.J. (2006) 'Semen displacement as a sperm competition strategy,' *Human Nature* 17: 253–64.

Mate infidelity and mate-retention tactics: Goetz, A.T., Shackleford, T.K., Weekes-Shackleford, V.A., Euler, H.A., Hoier, S., Schmitt, D.P. and LamUnyon, C. (2004) 'Mate retention, semen displacement, and human sperm competition: A preliminary investigation of tactics to prevent and correct female infidelity', *Personality and Individual Differences* 38: 749–63.

UWA sperm competition study: Simmons, L.W., Firman, R.C., Rhodes, G. and Peter, M. (2004) 'Human sperm competition: Testis size, sperm production and rates of extrapair copulations', *Animal Behavior* 68: 297–302.

14 From Clasper to Penis: We've Come a Long Way, Baby
Daniel Dennett: Dennett, D. (1995) *Darwin's Dangerous Idea: Evolution and the meaning of life*, Simon & Schuster, New York.
Peter Watson: Watson, P. (2000) *A Terrible Beauty: The people and ideas that shaped the modern mind*, Phoenix, London.
Hox gene Nobel prize research: Tallack, P. (ed.) (2003) *The Science Book*, Weidenfeld & Nicolson, UK.
Cliff Tabin and sonic hedgehog Hox genes: Shubin, N. (2008) *Your Inner Fish: A journey into the 3.5 billion year history of the human body*, Penguin, London; Dahn, R.D., Davis, M.C, Pappano, W.N. and Shubin, N.H. (2007) 'Sonic hedgehog function in chondrichthyan fins and the evolution of appendage patterning', *Nature* 445: 311–14.
Catherine Boisvert: www.armi.org.au/About_Us/Staff/Catherine_Anne_Boisvert.aspx.
Australian Regenerative Medicine Institute: www.armi.org.au/.
Genital developmental biology: Cohn, M. (2004) 'Developmental genetics of the external genitalia', *Advances in Experimental Medicine and Biology* 545: 149–57.
Hoxd13 and claspers: Freitas, R., Zhang, G. and Cohn, M. (2007) 'Biphasic Hoxd gene expression in shark paired fins reveals and ancient origin of the diatl limb domain', *PLoS ONE* 2(8): e754. doi:10.1371/journal.pone.0000754.
Lerista species: Skinner, A. and Lee M.S Y. (2009) 'Body-form evolution in the scincid lizard *Lerista* and the mode of macroevolutionary transitions', *Evolutionary Biology* 36: 292–300.

Epilogue
Information about Van Leeuwenhoek, Spallanzani and Hertwig from: Pinto-Correia, C. (1997) *The ovary of Eve: Egg and sperm and preformation*, University of Chicago Press, USA.
Thomas Aquinas's ideas: see Kenny, A. (2007) *Medieval Philosophy*, Oxford University Press, USA.

ACKNOWLEDGEMENTS

Firstly, for inspiring the work and for many helpful discussions and sharing of ideas, I sincerely thank my close colleagues and co-authors on our series of papers on early fish reproduction: Gavin Young, Kate Trinajstic, Tim Senden, Zerina Johanson and Per Erik Ahlberg.

The following people provided helpful information, quotes for the book, and in some cases directed me to further resources: Tim Birkhead, Aldo Poiani, Derek Siveter, Jason Dunlop and Hans Dieter-Sues.

For reading draft versions of chapters (or in some cases the entire book) and providing helpful feedback and comments I'm indebted to Luis Chiappe, Tim Flannery, Jared Diamond, Brian Brown, Robyn Williams, Carmelo Amelfi and Michael Shermer.

For help with images I thank Kevin McCracken, Jason Dunlop, Kate Trinajstic, Gavin Young, Tim Senden, Marty Cohn, Renata Freitas, Derek Siveter, Luis Chiappe, Rick Shine, Vanessa Woods, Dr Hans Pohl, and the Natural History Museum of Los Angeles County.

For supporting my research and field work at Gogo in

2005 and 2008 I thank Patrick Greene and Robin Hirst of Museum Victoria, Melbourne. For help in the field at Gogo during the 2005 trip that found the mother fish, I thank my colleagues Lindsay Hatcher, Mike Nossal, Brian Choo, Malcolm Carkeek and Tim Senden. Alf Kuhlman at Reel Images created the animations of our discoveries. Paul Willis provided helpful discourse during the field work at Mt Howitt and whilst filming features on the work for ABC TV.

For access to collections during our research I thank Zerina Johanson of the Natural History Museum, London; Lars Werdelin and Thomas Morrs of the Natural History Museum, Stockholm; and Mikael Siverson at the Western Australian Museum, Perth.

For permission to work in Paddys Valley and having the privilege of sharing that amazing landscape, I humbly thank the Gooniyandi people of Fitzroy Crossing, in particular Laurie and Rosalita Shaw, Maryanne and Louis Dolby. For permission to work on Gogo Station and for all his support of our work, I sincerely thank Dan Grant.

Much of the research leading to the discoveries of the mother fish and our other finds was supported by three Discovery Grants from the Australian Research

Council; support of field work at Gogo in 1986 and 1987 resulting in the discovery of the *Austroptyctodus* with three embryos, came from a major grant from the National Geographic Society.

I would like to thank my agent, Margaret Gee, for her continuous encouragement and faith in, and I also acknowledge support from my publisher, Jeanne Ryckmans, and the team at HarperCollins, Australia. Sincere thanks to the team at the University of Chicago Press for smooth production of this US edition of the book.

And, finally, I thank from the bottom of my heart my dear wife, Heather, who put up with and supported me through all the late nights and lost weekends of writing and researching, and who kindly edited and commented on the first draft of this book.

INDEX

The Dawn of the Deed

Index

Index

The Dawn of the Deed

Index

The Dawn of the Deed

Index

277

Dr John Long is one of the world's leading paleontologists and an award-winning author. Currently Vice President of Research and Collections at the Natural History Museum of Los Angeles County, John was previously Head of Sciences at Museum Victoria, and Curator of Vertebrate Palaeontology at the Western Australian Museum. John Long received the prestigious 2011 Royal Society of Victoria Research medal.